PHYSICS
WITHOUT MATHEMATICS

About the Author

Professor Clarence E. Bennett was head of the Department of Physics of the University of Maine from 1939 to 1967. Previously he was a member of the faculties of Brown University and of the Massachusetts Institute of Technology. He holds the Ph.D. degree from Brown University.

A fellow of the American Association for the Advancement of Science and of the American Physical Society and member of the American Association of Physics Teachers, the Optical Society of America, the American Society for Engineering Education, Phi Beta Kappa, Phi Kappa Phi, Tau Beta Pi, Sigma Xi, and Sigma Pi Sigma, he has contributed articles to scientific journals and is the author of the College Outlines *Physics* and *Physics Problems*, and of a standard textbook, *First-Year College Physics*.

PHYSICS
WITHOUT MATHEMATICS

Revised Edition

CLARENCE E. BENNETT
*Professor of Physics
University of Maine*

BARNES & NOBLE BOOKS
A DIVISION OF HARPER & ROW, PUBLISHERS
New York, Hagerstown, San Francisco, London

©
COPYRIGHT, 1949, 1970
By BARNES & NOBLE, INC.

All rights reserved. No part of this book may be reproduced or utilized in any form or by any means, electronic or mechanical, including photocopying or recording, or by any information storage and retrieval system, without permission in writing from the publisher.

L. C. Catalog Card Number: 76-124362

ISBN: 0-06-460067-X

84 85 86   10 9

Manufactured in the United States of America

PREFACE

That the nonscience student of today is interested in the physical world about him, and realizes the importance of scientific knowledge, need not be questioned. The many orientation and survey courses offered in our colleges and universities bear witness to this fact. It is doubtful, however, if the problem of introducing serious but nonprofessional science students to the field of physics has yet been adequately solved. The author of the present text has come to feel that the answer does not lie so much in eliminating most of the technical parts of the subject, as it does in divorcing the subject matter of physics as much as possible from arithmetic, for it is the numerical part of physcis at which the nonscience student shudders with the usual comment that he never could "see" mathematics. Furthermore, there is a surprising amount of material in a college physics course which need not be treated numerically. Experience has shown the author that a one-semester "nonarithmetical" course, which emphasizes concepts, vocabulary, and the definitions of technical terms, is not only possible but practicable for these students who will probably not go on in physics. It has been found that a real appreciation of the subject, not to be confused with a false sense of comprehension, can be developed in this manner.

To be sure, the arithmetic cannot be completely eliminated, but it can be minimized almost to the vanishing point. Of course this results in a corresponding sacrifice in the terseness and the rigor which properly characterize the general college physics course, but this need not be overdone. Although the language of physics is, to a certain extent, the language of mathematics, because the physicist appreciates this abbreviated and indispensable symbolic form of expression, it does not follow that the student who lacks a mathematical bent cannot appreciate a large part of that fundamental science which plays such an important part in the world in which he lives. Moreover, the physicist should feel an obligation to help his nonscientific colleague obtain a clearer picture of the world of physics and should be willing to forego the niceties of mathematical language when necessary. Such a feeling has motivated the preparation of this book.

In this text the order of development of the topics is a logical one designed to emphasize the unity of the subject matter. The relationships between each concept and the ones preceding it are carefully indicated so that the entire subject is developed from fundamental considerations. Technical vocabulary, however, is not avoided. It is emphasized, to encourage correct usage of technical terms. Only the arithmetic is minimized.

The present edition of this text is a complete revision of the earlier edition. In addition to extensive alteration and rearrangements of the material, a considerable amount of what is currently referred to as Modern Physics has been added. This has been presented from the point of view that such material can be appreciated properly only after many of the concepts and laws of classical physics are understood. It is not the intent of this book to mislead the nonscience student by exploiting the glamorous aspects of electronics and nucleonics. It is rather to help him develop the proper perspective toward the whole subject of physics, of which the nature of the atom is only a part.

It is definitely expected that a course based on this text will be a lecture course and that considerable attention will be given to lecture demonstrations. Physics, unlike many subjects, lends itself to the lecture type of presentation because of the almost unlimited amount of demonstration equipment and visual-aid material which is available. The lecturer will find innumerable opportunities to expand upon the material included in this text according to his available equipment. These pages are not intended to replace the spoken word, but rather to serve as a guide for reading assignments to supplement it, and as an outline of subject matter.

To facilitate the administration of such a course of one semester in length, the material is subdivided into chapters each of which may be considered roughly as a weekly assignment. Questions are listed after each chapter and general review questions are inserted at places appropriate for hour quizzes. The latter are of the "multiple choice" type; they can all be answered on the basis of this text, but they have intentionally been made difficult enough to require a genuine understanding of the points covered.

The student is cautioned against rapid and careless reading of this text. In the absence of arithmetic, words have to carry an extra burden, and they must be read carefully, even slowly, for their full meaning. Time must be allowed for reflection.

CONTENTS

CHAPTER		PAGE
I.	INTRODUCTION: Scientific Method, Fundamental Concepts, Measurements, and Basic Units	1
II.	HISTORICAL CONSIDERATIONS: Ancient and Medieval Physics, The New Awakening in Physics, Classical Physics, Modern Physics	12
III.	MECHANICAL CONSIDERATIONS: Forces and Motion, Vector Nature of Force, Equilibrium, Laws of Motion, Momentum	28
IV.	MECHANICAL CONSIDERATIONS (*Continued*): Work, Energy, and Friction	46
V.	ELASTIC CONSIDERATIONS: Elasticity, Vibrations, and Fluids	59
VI.	WAVES AND SOUND	72
VII.	MATERIAL CONSIDERATIONS: Constitution of Matter, Properties of Gases, Surface Effects	86
VIII.	THERMAL CONSIDERATIONS: The Nature of Heat, Thermometry, Expansion, Calorimetry, Change of State	100
IX.	THERMAL CONSIDERATIONS (*Continued*): The Nature of Heat Transfer, Quantum Theory, and Certain Other Philosophical Considerations	115
X.	ELECTRICAL CONSIDERATIONS: Static Electricity, Charges, Potential, Capacity	127
XI.	MAGNETISM	141
XII.	ELECTRICAL CONSIDERATIONS (*Continued*): Current Electricity	150

CHAPTER	PAGE
XIII. ELECTRONIC CONSIDERATIONS AND ATOMIC PHENOMENA: Introduction to Modern Physics, Electrical Discharges in Gases, Electronics, X Rays, Radioactivity, Nuclear and Solid State Physics	164
XIV. OPTICAL CONSIDERATIONS: Photometry, Laws of Geometric Optics, Optical Instruments	184
XV. OPTICAL CONSIDERATIONS (*Continued*): Physical Optics, Dispersion, Spectroscopy, Interference, Diffraction, Polarization	202
APPENDIX: Supplementary Review Questions	219
INDEX	225

CHAPTER I

INTRODUCTION

SCIENTIFIC METHOD, FUNDAMENTAL CONCEPTS, MEASUREMENTS AND BASIC UNITS

To many, the science of physics is a profound, highly mathematical, and exceedingly difficult subject to be undertaken only by those gifted with special mathematical abilities and to be avoided by all others. Although it is true that physics is probably the most highly organized branch of science today and does require profound mathematical analysis for its complete comprehension, it is also true that it is so closely related to almost every activity of everyday life that some familiarity with its laws and concepts should be had by every person who lays claim to being educated. Moreover, contrary to popular belief, a real insight into the nature of the physical world can be obtained without the use of highly mathematical language. Naturally the professional physicist is unwilling, if not unable, to dispense with such a useful tool of concise analytical expression as is afforded exclusively by mathematics; but the lay reader possessing a reasonable degree of scientific curiosity should not be deceived into thinking that the wonders of this fundamental science are forever denied him if, with an ability to reason logically, he is but willing to undertake the cultivation of a technical vocabulary. Of course, this is not a trivial undertaking and will require concentration.

Scope of Physics. A brief consideration of such matters as force, work, motion, energy, fluid phenomena, waves, sound, heat, electricity, magnetism, radio, atoms, electrons, lenses, prisms, optical instruments, color, and polarized light is, in general, the scope of this study. These topics are to be considered, however, not as a series of unrelated things, but rather as a part of a single subject —physics, the study of the physical world in which we live. All these topics are characterized by technical terms, and one of the purposes of this study is to develop a proper appreciation of these terms by learning their correct definitions as well as to acquire a familiarity with certain fundamental relationships which experience has shown to exist between them. In this way only can an appreciation of physics be obtained, when account is taken of the high degree of organization that the science has achieved.

Definition of Terms. Most of the terms used in physics are defined terms; that is to say, they take on their precise technical meaning merely because of common agreement. It will therefore be necessary at the very outset to recognize that the defining of new terms is not just a preliminary but is an indispensable condition for the success of our whole program. It is perhaps not an exaggeration to say that an introduction to physics is largely a study of one concept after another, each of which is carefully defined in terms of concepts more fundamental than itself. In a sense such an introduction is essentially a vocabulary study. Thus a structure of related concepts is developed, based upon certain ones which are considered fundamental and incapable of definition in terms of anything more basic. The full significance of this statement is not easily realized, but when the student finds that all the concepts of mechanics involve only three fundamental concepts—length, mass, and time—and that all the rest can be derived from these, he begins to appreciate the unity of physics. A consideration of these three fundamental concepts is to be the first detailed study in this text, but before this is undertaken, several broad aspects of the situation will be considered.

Inductive vs. Deductive Reasoning. The foregoing references to the logical development of this subject from fundamental considerations to more complex concepts may possibly convey the impression that physics exemplifies purely deductive reasoning. On the other hand, the very expression "scientific method" implies "inductive reasoning." It should be pointed out, however, that inductive reasoning characterizes the research phases of a science. After a body of information has been discovered and properly organized, it is often passed on most effectively to future generations of students by deductive processes. Deductive reasoning was developed to a high degree by the Greek philosophers, but deductive reasoning alone was unable to advance the study of physics beyond certain limits. False premises lead to wrong conclusions in spite of correct reasoning. Thus

Aristotle,[1] the famous Greek philosopher, reasoned that heavy bodies fall faster than light ones, and this conclusion was accepted as true for hundreds of years simply because of the authority behind it. It was not until the time of Galileo (1564–1642) that the matter was tested by experiment and found to be false. Galileo is often called the father of experimental physics because he was one of the first to advocate the basing of conclusions upon experimental fact rather than upon deductive reasoning alone.

The Scientific Method. The scientific method, as initiated by Galileo, is inductive and usually involves the several following distinct steps. In the first place *observations* are made. Then a *hypothesis* is postulated in terms of which the observed facts are consistent. A successful hypothesis becomes the basis of a theory which not only "explains" but suggests further observation in the form of *experimentation*. Finally a successful theory leads to the establishment of a *scientific law*. Thus science is advanced by observation and inductive reasoning, but it would be incorrect to conclude that there is a single scientific method. Scientists actually do not follow such a check-list procedure. Indeed they often resort to devious indirect procedures even to following their hunches, but in retrospect it usually appears that the above steps seem to have been followed, more or less. It was by such a procedure that the laws of motion of the heavenly bodies were "discovered." Attention is called particularly to the role played by observation in the process. Reasoning alone, no matter how correctly it is done, is not sufficient to establish a scientific law or principle. Observable facts are facts regardless of the methods used to explain them. Thus physics is essentially an experimental science; yet the student need not follow the same path as the pioneer in order to understand what the pioneer has accomplished. In this text the student will not be required to "discover" for himself but it is hoped that he will develop a real appreciation for what has been "discovered" by others.

Nature of Explanations. A word regarding the nature of explanation is perhaps in order at this point. An explanation of anything is always a relative matter. It is the making of a thing seem reasonable to a person in terms of his past experiences. Different degrees of appreciation have to be recognized, and what

---

[1] See chapter II for historical summary of developments in physics.

constitutes an explanation of something to an advanced student of a subject is likely to be quite unsatisfactory, if not unintelligible, to a beginner. Thus, as already pointed out, mathematics is essential for the explanation of physical phenomena for the more advanced student of the science, but to the uninitiated, explanations in terms of mathematics often become unsatisfactory. To the beginner, such matters as vocabulary are more important than mathematical relationships, although some of the concepts are admittedly mathematical by their very nature. Fortunately for this study, however, there is ample material for qualitative, or descriptive, considerations in elementary physics to enable the beginner to get a fairly good idea of the nature of the subject. Although the explanations and definitions in this text appear verbose if compared with the standards of the professional physicist, they are nevertheless adequate for our present purpose.

Quantitative Observations—Measurement. Since the scientific method not only calls for an ability to reason but also for careful observations, it follows that measurement plays an important role in physics. Observations must be quantitative as well as qualitative. Thus the question "how much" must always be answered before it is safe to base theoretical conclusions on any observation. This does not mean that the present reader must concern himself unduly with arithmetical details, but it does mean that the experimental physicist must give as much attention to his tools, which are all forms of measuring instruments, as the theoretical physicist gives to the laws and processes of mathematical analysis and logic. The distinction between experimental and theoretical (or mathematical) physicist is often made, but it is important to note that the theoretical physicist is limited to the material furnished by the experimentalist.

Direct vs. Indirect Measurements. Measurements of all kinds have to be made, but a remarkable thing about physics is the fact that not all measurements have to be made directly. In fact, many of the so-called measurements in physics, such as the determination of the size of an atom, or the measurement of the velocity of light, etc., are indirect measurements. This again is the result of the high degree of organization of the subject. The laws of physics and their relationships require that certain things must be so if certain other things are true. Thus two simple measurements of length yield a value of the area of a rectangle. Similarly, a measurement of length and one of time suffice to determine speed, to an accuracy comparable with that of the two measurements. Furthermore, a determination of length, one of time, and one of mass enable the physicist to measure the kinetic energy of a body, or its momentum, or any one of many quantities to be defined later. In brief, no more than these three kinds of direct measurement are required to determine any mechanical characteristic of a physical body. This is merely another way of stating our earlier assertion, viz., that all the concepts of mechanics are derived concepts expressible in terms of three fundamental ones: length, mass, and time. Thus we see that the direct measurements in physics, at least in the field of mechanics, can be restricted to measurements of these three fundamental things. These are commonly referred to as scale readings of length, pointer readings of mass, and clock readings of time.

Nature of Measurements. If we stop to consider, we realize that a measurement, such as a measurement of length, for example, is nothing but a comparison with a standard, which itself is nothing but a unit determined by common agreement. Thus a table is said to be three feet long merely because its length coincides exactly with that of a standard yardstick held along it. Similarly, the length of a rug, or a room, may be determined with respect to a yardstick, a foot rule, a measuring tape, or for that matter any convenient unit of length. Therefore length is something which can be measured in terms of units, but which otherwise has no significance except as it constitutes a basic concept for the development of other concepts. In other words, a basic concept such as length cannot be defined in terms of anything simpler, but a measurement of it can be made in terms of a standardized unit. Similarly, mass and time are quantities which are measured by comparison with standard units of these quantities. It should therefore be quite clear that the first task in the study of physics is to understand the manner of specification of the units in terms of which length, mass, and time are measured, for everything else seems to depend upon these units.

Incidentally, it should be noted that it is only by custom that these three particular concepts are taken as basic for the development of the entire subject. Theoretically any three concepts could have been used, but these three give the most natural, and probably the simplest, development.

Without further delay we shall proceed to the specification of the generally accepted units of length, mass, and time, with the understanding that all the concepts of physics are to be expressible in terms of units, each of which will be related in some way to the units of these three quantities. Complicating the situation somewhat, we find that agreement is not universal in these matters and that at least two independent systems of units must be considered, the so-called metric system and the English system of units.

The English standard of length is the *yard*, which is the distance between two scratches on a certain bar carefully kept in London. This length is subdivisible into three feet or thirty-six inches. The working unit in this system is taken as the *foot*. The metric standard of length is the *international meter*, originally defined as the distance between two scratches on a certain platinum-iridium bar preserved near Paris, France, but more recently (1960) redefined in terms of wave lengths of light given off by a certain isotope of the element krypton when excited by an electrical discharge. The terms wave length and isotope will be defined later. The number of such wave lengths is actually 1,650,763.7 for the standard meter. Although this unit was originally intended to represent a certain fraction of the earth's circumference, continued measurements of the latter quantity have disclosed errors in the original determination; but this does not in any way affect the agreement by which this particular length is accepted as the standard. Secondary standards, which are replicas of the original, are kept in every country. The one for this country is preserved at the National Bureau of Standards in Washington, D. C. Incidentally, the American standard of length is the American yard, which, although approximately equal in length to the English yard, is legally defined as 3600/3937 of the international meter. One one-hundredth part of the meter is the *centimeter*, which has been recognized for a long time as the working standard of length in scientific work. Thus, the metric and English units of length are connected by the approximate relationship that one inch equals 2.54 centimeters. This is worth remembering. It is also worth noting that international commissions are constantly at work maintaining and improving the legal status of units for purposes of trade. Indeed, a large portion of the program of the National Bureau of Standards is devoted to just such matters.

Units of Mass. To understand next how the unit of mass is defined, one must first be sure that the concept of mass is clear. It is not such a familiar concept as length. *Mass* is defined as a measure of inertia, where *inertia* is, by definition, simply a certain property of matter, by virtue of which a body tends to resist any change in its motion. The state of rest is considered a special case of motion, namely zero motion. It will be pointed out later that a force is necessary to change the motion of a body, whence it follows that inertia is a property involving force and motion. For the purpose of this discussion, however, a *force* is to be thought of simply as a push or a pull tending to change the motion of a body.

Weight Not the Same as Mass. One of the commonest forces to act upon a body is the pull of gravity, that is to say, the attraction of the earth. All bodies on or near the earth are under such an influence at all times. Technically the *weight* of a body is the pull of gravity on it. See fig. 1.1. Thus a man is said to weigh 150 pounds because the earth attracts him with a force of 150 pounds. Even in the spaceship the earth attracts the astronaut. It should be pointed out that the condition of weightlessness which he experiences is somewhat of a misnomer; more correctly he should be referred to as being *apparently* weightless with respect to the spaceship. The earth, however, still pulls him and the spaceship together to keep them in their almost circular orbit, but with decreasing magnitude as the distance from the earth increases. In their flight to the moon, the astronauts reached a distance from the earth at which the pull of the earth exactly balanced the pull of the moon, beyond which, and on the surface of the moon, the pull of the latter completely obscured the pull of the earth. Thus moon-weight is one thing and earth-weight is another, the latter being approximately six times the former.

Fig. 1.1. The weight of a body can be determined by a spring balance.

Attention is called to the fact that the common and familiar unit of force is the *pound*. In the metric system the units of force are the *dyne*, roughly equivalent to the weight of a thousandth part of a postage stamp, and the *newton*, which is the weight of the standard kilogram. The pound is sometimes used as a unit of mass, a concept quite different from weight, and confusion arises from the use of the same unit to express two different concepts. Such difficulty can be avoided only by painstaking care in the use of technical language. See pp. 39–41 for a more complete discussion of this matter.

It is obvious that a body without mass would be weightless (indeed, it would not exist), yet mass and weight are different concepts. Mass, however, is considered by the physicist to be the more basic concept of the two, and the unit of mass is arbitrarily defined. A certain piece of platinum, *having just this characteristic by definition*, is preserved in London along with the English standard yard, and its mass is defined as a pound mass. See fig. 1.2. The United States pound is legally defined as a certain fraction of the mass of the international kilogram (see below). The standard pound of weight is now defined as the pull of gravity at a specific place (sea level and latitude 45°) on this one-pound mass. It is a fact that the pull of gravity on a body (its weight) varies slightly from point to point on the earth's surface, being greater at the poles than at the equator because at the poles it is nearest to the center of the earth. But at sea level and in latitude 45° (as London) the standard unit of mass and the standard unit of force are thus simultaneously defined in terms of a specific piece of platinum.

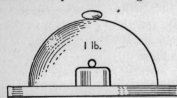

Fig. 1.2. The standard pound mass is preserved under a glass cover in London.

In a similar manner, another particular piece of platinum preserved near Paris is said to have a mass of one *kilogram*. One one-thousandth of this mass is called a *gram*, a unit which is generally accepted as the working unit of mass in the metric system, although the kilogram is the fundamental standard.

The metric system is a decimal system and has been accepted almost universally in scientific work because of its convenience. Engineers, however, both in Great Britain and the United States lean toward the English or the so-called British Engineering system of units, on the ground that it is more familiar to most people. This difference might well be used to illustrate an important distinction between science and engineering. Whereas science deals for the most part with truths systematically organized into a body of knowledge, with logic carried right down to the matter of units, engineering deals more with the applications of science toward commercially practicable ends. Compromises are frequently made with accepted customs and usages by practical people. Of course in the last analysis, there is much overlapping, and no one should attempt to distinguish too sharply between science and engineering, especially since engineering is increasingly becoming more and more scientific all the time.

Unit of Time. The last in this discussion of the fundamental units is the unit of time. Whereas time is often defined as a measure of duration, duration can only be defined as a measure of time. Thus time is a basic concept, not to be defined but rather to be measured in units. It has become almost universal practice to measure time by the use of clocks which have been calibrated to read in terms of the time required for the earth to rotate about its axis in its orbit about the sun. The mean solar day is the accepted unit, of which the mean solar second is a 1/86400 part. The length of the mean solar day is found by averaging over the whole year. This unit is used in both metric and English systems, and thus does not share in the confusion created by the other units. See fig. 1.3. In 1967 a new standard of time based upon one of the natural vibration frequencies of a certain isotope of cesium was adopted. This makes the second equivalent to the time required to make 9,192,637,770 such vibrations.

Fig. 1.3. Intervals of time are measured by clocks.

Systems of Units. In brief, then, we have outlined the basic units in three systems. The particular metric system in which length is measured in centimeters, mass in grams, and time in seconds is called the C.G.S. (centimeter, gram, second) system. At the present time an attempt is being made to popularize the use of a metric system called the M.K.S. system in which the meter is actually used, not just defined, as the fundamental unit of length, and the kilogram is actually used instead of the gram as the fundamental unit of mass. This system is a compromise between scientists and engineers, and its merits lie chiefly in the electrical units which come later in this study. In this text all three systems will be used, but when references are made to engineering applications of science for purposes of illustration, the English units will probably be used. Where English units are used, lengths will always be expressed in feet, force (not mass) in pounds (see chapter III for further considerations of mass and force), and time in seconds. In terms of such units the physicist records the direct measurements upon which his conclusions are reached and tested.

Measuring Instruments. It is seen that, in the last analysis, direct measurements boil down to scale readings of length,

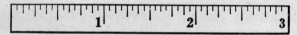

Fig. 1.4. Intervals of length are ordinarily measured by foot rules, yardsticks, meter sticks, etc.

clock readings of time, and pointer readings of mass. See figs. 1.4 and 1.5. Such pointer readings of mass are obtained with a balance by the use of which the tendency of a body placed in one pan, to be pulled downward by gravity, is balanced by supplying an equivalent amount of standardized mass in an adjoining pan connected to the former by a bar resting on a knife

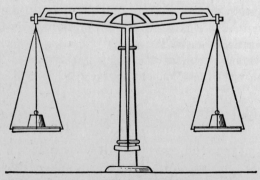

Fig. 1.5. Mass is measured by the beam balance.

edge at its center. In the course of his work the physicist has invented ingenious devices to facilitate these measurements, including the vernier and the micrometer calipers (see figs. 1.6 and

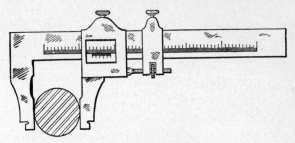

Fig. 1.6. The vernier caliper is used to make accurate measurements of relatively short lengths.

1.7), the measuring microscope, specialized forms of mass balances, and intricate clockwork mechanisms. Fundamentally, however, scale readings of length, mass, and time constitute all the direct quantitative observations.

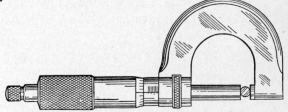

Fig. 1.7. The micrometer caliper is used to make accurate measurements of very short lengths.

Relative Value of Quantitative Considerations. Now, lest the beginning student should get the impression that this discussion of fundamental units appears to emphasize the numerical aspect of the subject in spite of promises that this book would minimize quantitative considerations, he should be reminded that much of the beauty of the science of physics to the physicist lies in its high degree of organization and the discipline which its strict definitions provide. In studying such a subject, even if the quantitative or numerical aspects are not pursued to the limit, it is well worth noting the importance to the physicist of getting first things first. Only thus does one get the correct orientation properly to appreciate the more complicated, and perhaps more fascinating, aspects of the story which is to be unfolded in the subsequent chapters of this book.

Summary. In summarizing this chapter we see that physics is to be considered as a logically organized body of concepts to be understood in terms which are uniquely defined. By inductive reasoning, based upon observation, conclusions are drawn regarding these concepts. These conclusions must be subjected to tests of further experimentation and observation. Although inductive reasoning is paramount to the research phases of the science—as true today as ever since the earliest times—deductive methods are used to advantage in passing the results on from one generation to another.

In this chapter the importance of measurement has also been emphasized because observation must be quantitative as well as qualitative. The physicist uses measuring instruments as well as instruments of the mind in making his observations, and expresses his results in terms of suitable units, of which there are a few more basic than the rest.

Future chapters will deal with some of the results of measurement, and although measurement naturally suggests numbers, the latter will be minimized so as not to interfere with descriptions of physical concepts. Before going on immediately to a detailed discussion of physical concepts, however, attention will be given in the next chapter to the place of physics in our civilization by a brief treatment of the history of the science of physics. Such a historical summary will constitute something of an over-all look at the field before a detailed consideration of the various parts is undertaken. It will also be a chronological rather than a logical development.

QUESTIONS

1. What is meant by the "Scientific Method"? Is it a specific method?
2. Distinguish between the meter, yard, foot, centimeter, millimeter.
3. For what purpose is the vernier caliper used?
4. What is the difference between "inductive" and "deductive" reasoning?
5. Distinguish between the newton, pound, and dyne.

CHAPTER II

HISTORICAL CONSIDERATIONS

ANCIENT AND MEDIEVAL PHYSICS, THE NEW AWAKENING IN PHYSICS, CLASSICAL PHYSICS, MODERN PHYSICS

In a descriptive survey of physics it is appropriate to consider, before getting absorbed in the detailed concepts of the subject, matters of a broad nature, such as historical developments. By such a procedure the student is provided an opportunity not only to appreciate how the science has grown, but also to evaluate its importance in, and to, modern civilization. After all, this is a scientific age and it is indeed a matter of some concern to the serious student to find out just what physics has contributed to his world and, for that matter, what his world has contributed to physics. This point becomes all the more significant when it is noted that physics is one of the oldest and one of the most highly organized of all the sciences, whose development has been very much affected by the political and economic conditions of people. Note that the atomic bomb was the result of a war situation and was produced by a country capable of spending over 2 billion dollars on the effort whereas the science which made it possible for man to land on the moon reputedly cost more than 20 billion dollars. The scientific discoveries which made these feats possible would never have been made by an economically backward society. Nor would the former have been motivated except for the pressures of a political war.

Certainly the contribution of physics to the cultural and economic welfare of mankind has been very great indeed in times of both peace and war throughout the ages. It is the purpose of the present chapter to outline some of the important developments in the science and to associate them with persons. Too often the student of science or technology overlooks the fact that physics is just as much concerned with people as it is with things. Discoveries of natural laws are made by persons whose own lives as well as those of others are affected by them. It is therefore appropriate that this historical outline be developed around the names and dates of famous physicists. The student should try to associate the dates with world events with which he may be familiar outside the realm of science. Possibly some connections between events of great social import and discoveries in science will be noted that have not previously attracted his attention.

Four Periods of Physics. It seems natural to consider this historical survey in four main subdivisions, which we shall refer to as periods of physics, as follows: *Ancient and Medieval Physics* (roughly 3000 B.C. to 1500 A.D.); *The New Awakening in Physics* (1500 to 1700); *Classical Physics* (1700 to approximately 1890) *Modern Physics* (1890 to the present time). These dates can be considered only as very approximate. The first and last of these periods might be subdivided further. A distinction could well be made between the ancient and the medieval periods, but the contributions made in the latter period, which might be said to extend from 500 A.D. to 1500 A.D., were so few as to make such a distinction hardly worth while. Moreover, the advances made in the so-called modern period have been so numerous that a break around 1925 might seem justifiable. Perhaps the advent of the atomic bomb in 1945 will make historians of the future consider this year the beginning of a new era. However, the present is not the time to evaluate the progress of contemporary physics and so it has seemed best in this survey to group all the accomplishments since 1890 into a single so-called modern period.

First Period. Ancient and Medieval Physics (3000 B.C.–1500 A.D.). Records indicate that the ancient Babylonians and Egyptians as early as 3000 B.C. were familiar with some of the fundamental principles of physics, especially those dealing with the measurement of land. There is also evidence that the beginnings of astronomy were somewhat understood by these early people. But nothing very definite, in a scientific sense, seems to have come from this. Of course we recognize the engineering achievement represented by the pyramids of Egypt, but the astronomy was more the sort of thing now known as astrology. It also seems certain that the Chinese were familiar with the magnetic compass earlier than 1000 B.C., but here again the records are very scanty.

ANCIENT AND MEDIEVAL PERIOD

THE GREEK PART OF THE FIRST PERIOD. From approximately 700 B.C. to 150 A.D. very definite advances in physics were made by the Greeks. *Thales* (640?–546 B.C.), a rather shadowy figure, appears to have recognized certain aspects of what today is known as static electricity. Historical scientists also credit him, and a group associated with him known as the Ionian School, with the crystallization of the view that *Fire, Water,* and *Earth* were fundamental substances. This certainly represented a recognizable beginning of scientific thought.

Other early names encountered in the records are *Pythagoras* (580–500 B.C.) and *Democritus* (460–370 B.C.). The former is remembered as the founder of the Pythagorean school of philosophers; his most important contribution was probably the Pythagorean theorem of geometry, which states that the hypotenuse of a right triangle equals the square root of the sum of the squares of the sides. Democritus is credited with postulating an atomistic view of matter. Although an atomistic view is held today, Democritus is not considered its father since there appears to have been little if any scientific basis for his postulate. It is more appropriately referred to as idle speculation than as a scientific postulate.

Next we come to *Plato* (427?–347 B.C.) and his pupil *Aristotle* (384–322 B.C.). Plato was undoubtedly one of the greatest of the Greek philosophers, yet his contributions to physics were relatively meager. His famous pupil Aristotle, however, was the great organizer of scientific knowledge of his time. He is credited with having written the first textbook in physics. He also contributed so much to the science of physics that for many centuries his influence was paramount. In spite of his greatness, however, we criticize Aristotle today on the ground that his famous theories were not based primarily upon experimental evidence. Although he frequently refers to the value of observed facts in the development of a theory, he seems not to have followed his own advice very conscientiously. Much of his work is based upon abstract argument not substantiated by observation. It is often stated that Aristotle was more concerned with explaining "why" things were than "how" nature behaved. Modern science, of course, is based on the latter point of view with considerable emphasis upon quantitative measurements made as nearly exact as possible. By present-day standards Aristotle was certainly the "armchair" type of philosopher whose conclusions were more deductive than inductive. Although observation was not entirely missing, it was more of the passive curiosity type not accompanied by purposeful experimentation. Nevertheless, Aristotle's name was unquestionably one of the greatest, if not the greatest, in the Greek period of science.

Another Greek of importance was *Euclid* (450?–374 B.C.). Although he made contributions to geometric optics, his fame is much greater in the field of mathematics than in physics.

Finally, in the Greek period, we encounter *Archimedes* (287?–212 B.C.) of the Alexandrian school. On the practical side Archimedes is perhaps the best known of all the Greek physicists. Today he would be called an engineer or an engineering physicist, by virtue of his many applications of physical principles to wartime as well as peacetime pursuits. Yet there are also those who would call him the founder of mathematical physics because of his apt and accurate applications of mathematics. His work in hydrostatics alone would make him one of the most famous of the ancients, but he made notable contributions to optics, mechanics, hydraulics, and other fields of physics as well. Unlike Aristotle, however, he interested himself in practical matters, or, as would be pointed out today, he kept his feet on the ground.

In summarizing the Greek period as a whole it seems fair to state that the entire period was characterized by deductive rather than inductive reasoning, that observation was for the most part superficial rather than critical, that there was more passive curiosity than active experimentation and more armchair speculation than fact finding. Today one cannot help wondering why this was all so. Yet it must be remembered that experimentation in physics is a manual procedure and that the social status of the free-born Greek citizen above the slave forbade manual labor. Whereas the impact of science on society is ordinarily acknowledged, we note here a definite impact of society on science, for it is undoubtedly true that had manual activity not been frowned upon in this period, more advancement would have been made by the Greeks. Instead, the science of physics proceeded just so far and no farther in a period of several hundred years compared with the enormous advances made in just a few years in our time.

THE LATTER PART OF THE FIRST PERIOD. Although the first period continued until about 1500 A.D., practically no advance in physics was made between 50 B.C. and 1550 A.D. This period covers the rise and fall of the Roman Empire, coinciding with the decline of Greek culture generally, and the overrunning of the ancient western world by the barbarians. The Romans gradually absorbed Greek culture, but by 600 A.D. all Europe had been deprived of the opportunity to avail itself of it. The Romans were not scientifically inclined. Practically all Greek manuscripts went to Arabia. In a way, Greek science was thus preserved for posterity by the Arabs, who themselves added very little to it. They did, however, introduce to science the so-called Arabic system of numbers. *Alhazen* (965?–?1039) produced a work on optics, to be sure, but generally speaking Greek science was not improved upon to any appreciable extent by its translation into Arabic. It was still based upon the authority of Aristotle.

Between 700 and 1100 A.D. a beginning was made toward a revival of learning in Europe. Large universities developed under the shelter of the church. Trade spread, and both Greek and Arabian manuscripts gradually found their way back to Europe. The crusades assisted in this process. Since the church had survived the Roman state and had become all-powerful, it was natural that the revival of learning, and of science in particular, should take place under its influence. Many of the scientific manuscripts were translated from the original Greek into Latin by the monks in the monasteries, where merchants and knights bringing treasures from the east, including the writings of Aristotle, would often seek shelter from the attacks of wandering bands of pirates and outlaws. These scholars apparently were satisfied just to make exact translations and so the science which they passed on to the world through the church was the original Aristotelian version unadulterated by any experimentation of their own. Thus the authority of Aristotle was carefully preserved. In this period the scholars stuck to the Greek and Roman numbers, leaving the Arabic numbers to trade, and so even the Arabian influence was minimized. This is another reason why it becomes possible to state that for approximately 1500 years there were practically no advances over Greek science. It was certainly a period of scientific stagnation. By the year 1500 or so, science had just about returned to the status which it had occupied 1500 years earlier. Nevertheless, the church had reestablished science in the various large universities, directly under its control. It is easy to see, however, that church domination flavored it to suit itself. The doctrines of Aristotle came to have the power of law behind them. Truth was not discoverable, by that time; it was dictated by the church, and it became a crime of the first order even to question the church-sponsored views of Aristotle, to say nothing of suggesting that experimentation might be a better way to establish the truth. Naturally science degenerated under this system, in which it became so interwoven with ecclesiastical doctrine that scientific facts could not be separated from religious dictates. The time was not quite ripe for the new era which was to dawn about 1550.

Second Period. The New Awakening in Physics (1500–1700). As time elapsed and trade developed, wealth increased. This led to the development by wealthy merchants of universities for study and research outside the church. This movement nurtured a growing dissatisfaction with authority in science. By 1550 doubters of Aristotle appeared with experimental proof of their views. That is to say, doubters began to express themselves openly. As a matter of fact there had probably always been doubters but they had been successfully suppressed. Two centuries earlier *Roger Bacon* (1214?–1294) had taught that belief should be based on observation and experimentation rather than on authority, but such expressions resulted in his spending practically the last third of his life in prison. Truly he was a man who lived at least two centuries ahead of his time.

Also, in Italy, there was *Leonardo da Vinci* (1452–1519). But in spite of the great accomplishments of this man in practically every field of the arts and sciences, he was hardly known in his time and his influence was practically nil. He left his works in manuscript form and they have come to be appreciated only in recent times. Now it is realized that he actually was one of the great scientists of all time.

The period of new awakening really commences with *Copernicus* (1473–1543), *Galileo* (1564–1642), *Tycho Brahe* (1546–1601), *Kepler* (1571–1630), and *Gilbert* (1540–1603), who all paved the way for the great *Isaac Newton* (1642–1727). Copernicus developed the heliocentric theory of the universe. Galileo,

Tycho Brahe, and Kepler established the fundamental ideas of modern celestial mechanics, based upon observation first and theorizing afterward, thus revolutionizing scientific thought. Galileo in particular stressed the idea of controlled experimentation to such a degree that today he is recognized as the rather of the modern scientific method based upon inductive rather than deductive learning. However, he, as well as some of the others, paid the price of propagating revolutionary ideas by spending periods of his life in prison.

Galileo carried observation to the quantitative stage by making accurate measurements. He truly emphasized the "how" (even the "how much") as contrasted with the "why" of Aristotle. By his quantitative observations on falling bodies and other mechanical motions, assisted by instruments of his own invention to improve the accuracy of his measurements, he laid the foundation for the discoveries of Newton, who was born the year he died. His work was not limited to the field of mechanics, however. His contributions to other fields, including optics, were numerous.

Sir Isaac Newton is considered by many to be the greatest scientific genius the world has produced. He crystallized the scientific thought of his time into a few fundamental statements now accepted as laws of nature. These include three famous laws of motion and the law of gravitation in the field of mechanics alone. In addition, he invented the calculus and contributed greatly to the field of optics. His role was primarily that of a co-ordinator of information or a systematizer of knowledge. He formulated the over-all pattern by which scientific knowledge was to be organized in the great classical period that was to follow his time and which has not yet become outmoded, although today it has been supplemented (not replaced) by new and sometimes conflicting ideas in the realm of atomic physics. By 1690 much of mechanics was known and fitted into the Newtonian pattern.

Newton's contemporaries include *Huygens* (1629–1695), who proposed the wave theory of light in opposition to Newton, and other famous physicists including *Boyle* (1627–1691), well known in connection with gas laws, and *Hooke* (1635–1703), whose work in elasticity is famous. *Pascal* (1623–1662) is remembered for his law of fluid pressure.

Thus the second great period in physics ends around 1700 or so, when the ideas of Galileo, Bacon, and others, who were aware of the inadequacies and the erroneous conclusions of Aristotle and were irked by actions of the church in forcing the same upon them as religious doctrine, finally took hold and opened up a vast new era. It was certainly a period of New Awakening and it paved the way for the physics of the eighteenth and nineteenth centuries, during which time new discoveries, although made in rapid succession, seemed to do hardly more than confirm the broad structure of the science laid down by Newton.

Third Period. Eighteenth and Nineteenth Century Classical or Newtonian Period (1700–1890).

It will not be feasible in this historical survey to enumerate all the achievements of the eighteenth and nineteenth centuries. The science of physics was really gathering momentum by this time and becoming very complicated. The science was getting large enough for the various subdivisions to be significant. Advances were made in the fields of mechanics, heat, light, and electricity, as if each branch were more or less independent, but the work of Newton provided the means of integrating all this knowledge. In a way the period can well be described as one in which the fundamental views of Newton gradually came to be appreciated and established. A wealth of quantitative material was gathered, but it all seemed to fit together in a most remarkable way into the Newtonian pattern. Indeed, the success of Newtonian physics was so great that by the end of this period it almost seemed that the end of physics was in sight. Almost everything seemed settled; hence the designation "classical period."

Rather than attempt to develop this period in a chronological order as in the case of the two preceding periods, we shall consider the various branches of the science separately and do hardly more than list the outstanding names, dates, and events in this period. As noted earlier, this is done in the hope that the non-science student may associate such names, dates, and scientific events with corresponding names, dates, and events of social or cultural significance. In what follows it should be realized that many of the words used and much of the technical terminology will be explained at appropriate places later in the text. In fact that is what this text is really all about.

CLASSICAL PERIOD

Commencing with mechanics, we note the work of *Bernoulli* (1700–1782) in hydrodynamics and gas theory, *D'Alembert* (1717?–1783), *Euler* (1707–1783), *Lagrange* (1736–1813), and *Laplace* (1749–1827) in theoretical mechanics.

In the field of heat, the period 1600 to 1800 saw the development of thermometers and temperature scales by Galileo, *Fahrenheit* (1686–1736), and others. It also saw the introduction of the concepts of latent and specific heats by *Black* (1728–1799) and the development of the steam engine by *Watt* (1736–1819). The nineteenth century saw the concepts of heat fitted into the Newtonian picture by the work of *Rumford* (1753–1814), *Joule* (1818–1889), and *Rowland* (1848–1901). These men established the view that heat was merely a form of energy. *Carnot* (1796–1832), *Mayer* (1814–1878), *Helmholtz* (1821–1894), *Kelvin* (1824–1907), *Clausius* (1822–1888), and others established the fundamental laws of thermodynamics, in which the basic concept of energy served to unify the concepts of heat with those of mechanics. Finally we note the work of *Gibbs* (1839–1903) in chemical thermodynamics and later in statistical mechanics closely related to the subject of heat.

Also in the field of light, many advances were made in these two centuries, and again the unifying influence of the Newtonian methods was very apparent in the latter one. Galileo had attempted to measure the velocity of light but had found it too great for accurate determination. *Roemer* (1644–1710) and *Bradley* (1693–1762) made more successful determinations, arriving at values not too far different from that accepted today—around 186,000 miles per second. *Foucault* (1819–1868) and *Fitzeau* (1819–1896) refined the measurements around 1850, but the final result waited for *Michelson* (1852–1931).

Huygen's wave postulates were revived by the work of *Young* (1773–1829) and *Fresnel* (1788–1827). *Malus* (1775–1812) discovered the phenomenon of polarization of light by reflection. This work was extended by *Brewster* (1781–1868). *Fraunhofer* (1787–1826), *Kirchhoff* (1824–1887), and *Bunsen* (1811–1899) laid the foundation for modern spectroscopy.

The crowning climax of this period in the field of light was the supposed stroke of finality given to the age-old question of the nature of light, whether corpuscular or wavelike, by the great *Maxwell* (1831–1879), whose electromagnetic wave theory of light satisfied all parties to the controversy. As a matter of fact, Maxwell is considered by many to be the greatest theoretical physicist of the nineteenth century as a result of his extraordinary accomplishments in the fields of light and electricity.

The last field to be considered in this third period of physics is that of electricity, or electricity and magnetism, for it was found in this period that these two fields were really one. Probably more attention was given to electricity than to any other branch of physics during the eighteenth century. *Gray* (1670–1736), *Du Fay* (1698–1739), *Franklin* (1706–1790), *Cavendish* (1731–1810), and *Coulomb* (1736–1806) did significant work in electrostatics. *Galvani* (1737–1798) and *Volta* (1745–1827) were pioneers in current electricity. In the latter part of the period the outstanding contributors were *Faraday* (1791–1867), *Oersted* (1777–1851), *Ohm* (1789–1854), *Henry* (1797–1878), and *Maxwell* (1831–1879), although they had many contemporaries, including *Ampère* (1775–1836), *Wheatstone* (1802–1875), *Lenz* (1804–1865), *Kelvin* (1824–1907), *Kirchhoff* (1824–1887), and *Hertz* (1857–1894). Faraday was the outstanding experimentalist of his time, and Maxwell, as stated earlier, was probably the most outstanding theoretical physicist of the period. In this field, as in all the others, the period ended with a great feeling of satisfaction by physicists that the science of physics was finally organized and very well integrated. Remarkable generalization had been accomplished by the theoretical physicists, especially Maxwell, utilizing the general methods of Newton and the highly successful concept of energy. It was indeed the classical period in physics. So complete was the picture and so satisfied were the physicists in their accomplishments that by 1890 a general feeling became widespread, although it is difficult to determine its precise origin, that physics was about done, that probably no startling new discoveries would ever be made, and that probably future generations of physicists would have to content themselves with merely extending the accuracy of known information, perhaps to the next decimal place. That this was not the case is the story of modern physics, which period started just about 1890 when physicists were jolted rather abruptly out of their lethargy by several noteworthy discoveries including those of the electron, the X ray, and radioactivity.

The Fourth Period. Modern Physics (1890 to the Present).
The great generalizations and correlations in theoretical physics and the refinements of measurement in experimental physics during the nineteenth century, especially the latter part, certainly placed the science in a strategic position with respect to the technological and industrial activity just commencing. The economic world was surely destined to feel the impact of it. Indeed, not even today has industry come anywhere near to exhausting the applications of classical physics to its needs, even though the fruits of a still more recent period, the modern period, have come to occupy front place on the technological scene. Of course we refer to electronics, television, atomic energy, etc., all of which were unheard of much before 1890. But it is still classical or Newtonian physics which properly constitutes the major part of a college program in physics today, especially for those students anticipating a career in engineering. This is because the so-called newer physics, which is the material covered in the fourth period, cannot properly be understood and evaluated without a substantial background in classical physics. Consequently this text will of necessity be largely classical in its scope, but not to the exclusion of atomic and nuclear concepts.

Although the achievements of the nineteenth century were outstanding, the impression must not be given that perfection had been attained. There were gaps here and there, and there were certain inconsistencies. For example, the electromagnetic wave theory of light was wonderful, but it required the postulate of a medium, the luminiferous ether, for the waves to wave in. Unfortunately, no one could find any tangible evidence for the existence of such a medium. Also the study of heat radiation was not quite settled. Furthermore, recent discoveries in electricity, such as the photoelectric effect by which the action of light produces an electric current, were inconsistent with Maxwell's theory of light. Yet all such difficulties were considered to be of very minor importance compared with the over-all success of the general Newtonian pattern. It was felt that sooner or later they would all be straightened out, like a jigsaw puzzle complete except for the last few pieces. The fact is, however, that instead of this happening the difficulties became enormously magnified as physicists concentrated their attention on them. It finally took a whole new point of view, such as is utilized in the quantum theory and in relativity, to straighten things out. But this did not happen until a series of brand-new discoveries were made, including the isolation of the electron, the discovery of radioactivity and X rays, and the formulation of the concept of the electrical structure of matter, all of which were likewise explained by the new viewpoint.

Crookes (1832–1919) had discovered cathode rays in 1878, but *J. J. Thomson* (1856–1940) explained their nature as negatively charged electrical particles, called electrons. This was between 1890 and 1897 and is usually referred to as the beginning of the electronic era, although *Stoney* (1826–1911) had suggested the name "electron" as early as 1874. *Lorentz* (1853–1928) formulated an electron theory of matter in 1895 in which vibrating electrons accounted for the electromagnetic radiations predicted earlier by Maxwell and discovered in 1888 by *Hertz* (1857–1894).

Also in 1895 *Roentgen* (1845–1923) discovered X rays, and in 1896 *Becquerel* (1852–1908) discovered radioactivity. These were followed in close succession about 1898 by the isolation of polonium and radium by the Curies, *P. Curie* (1859–1906) and *M. Curie* (1867–1934).

The year 1900 is marked by the introduction of the quantum theory by *Planck* (1858–1947). This, followed by the theory of relativity in 1905 by *Einstein* (1879–1955), set the stage for many advances because of the new way these two theories provided for viewing natural phenomena. These theories struck at the fundamental philosophy of physics and provided the key to the solution of several problems left over from the preceding period. The quantum theory explained heat radiation and the photoelectric effect almost immediately. The theory of relativity went farther than Newtonian mechanics to explain certain phenomena associated with matter moving with extraordinarily large velocity, such as electrons, atoms, and molecules. It also provided an explanation for the failure of the Michelson-Morely experiment (1887) to determine the drift speed of the earth through the luminiferous ether. Although these theories were very reluctantly received, they have nevertheless gradually become established in physics, in spite of the fact that they have so altered scientific thinking that many physicists of only a couple of generations ago confessed to their inability to reconcile some of the postulates with classical physics or even with common sense. Indeed, much of the newer theoretical physics is abstract mathematics, but most modern physicists have come to accept the results of these two theories as reasonable.

From about 1911 much interest has been focused upon the general field of physics known as atomic physics. *Rutherford* (1871–1937), with the help of many assistants over a period of years, finally established the concept of the nuclear atom. *Bohr* (1885–1962) devised a model of the atom in 1913 which was patterned after the solar system and its planets. This so-called planetary atom, consisting of a positively charged nucleus encircled by negatively charged electrons, has been very widely publicized. Although today this picture has been more or less replaced by abstract mathematical representation in the mind of the theoretical physicist, many of the Bohr features are still useful, especially in elementary explanations of atomic phenomena.

To the uninitiated this last statement may seem strange, raising as it probably does the question of how a picture can be accepted if it is not correct. This is an example, however, of the modern point of view in physics, and its relation to common sense. Quantum and relativistic physics have prepared the scientific mind to appreciate how complex nature really is, and how far the world of atoms and electrons is removed from that of common sense anyway. It reminds us that explanations are only relative to the student's background To an elementary student something can be a perfectly satisfactory explanation which would not be appropriate for the advanced student. In other words, there can be any number of ways of explaining anything. It is not a question of which is the correct one; it is rather a question of which is the better one for the purpose, i.e., which explains the most with the fewest assumptions. Thus perhaps it is clear how the newer physics has seemed so confusing to the older generation.

Many names might be mentioned in connection with atomic physics, but in this brief summary only a few seem appropriate. *W. H. Bragg* (1862–1942), *Aston* (1877–1945), *C. T. R. Wilson*

(1869–1959), and *Millikan* (1868–1953) were physicists who attained prominence before 1920. Of course no distinction can be made here between atomic and electronic physics, and furthermore these lists are very far from complete.

During the 1920's further advances were made in electronics, spectroscopy, and nuclear physics, although research in all fields increased greatly after the First World War. *A. H. Compton* (1892–1962) practically removed the last traces of doubt of the quantum theory in 1923 by his experiments with X rays. *De Broglie* (1892–   ) introduced the concepts of wave mechanics in 1924. *Heisenberg* (1901–   ), *Dirac* (1902–   ), and *Schrödinger* (1887–1961) did much to develop this field between 1921 and 1926. *Davisson* (1881–1958) in this country and *G. P. Thomson* (1892–   ) in England produced experimental evidence for the wave nature of the electron in 1927 and 1928.

In 1932, *Chadwick* (1891–   ) discovered the neutron, a fundamental particle which carries no electrical charge, and *Anderson* (1905–   ) discovered the positive electron, sometimes called the positron. *Lawrence* (1901–1958) invented the cyclotron in this same year. *Joliot* (1900–   ) and his wife *Irène Curie-Joliot* (1897–1956) discovered artificial radioactivity in 1934. *Fermi* (1901–1954) and others produced artificial radioactivity by neutron capture. *Hahn* (1879–   ) in 1938 discovered uranium fission, which is the basis of the atom bomb of 1945.

Thus we have touched upon some of the high points of physics right up to the present time. As stated before, the list of names and dates is far from complete, nor is it intended that the elementary student should fully appreciate all of the above at this stage of the study, but enough has been included to maintain the continuity. It is hoped that the reader will wish to refer back to this chapter from time to time as the logical developments of the concepts are described in subsequent chapters of the text. Research today, however, is not a matter for individual physicists. This is the age of group research, and in many of the projects noted above only the leader or leaders have been mentioned. Hundreds of others have contributed who may never receive full recognition for their work.

SUMMARY

It has been the aim of this chapter to outline the advances in physics from almost prehistoric times up to the present. It should be obvious that these advances have been made at an accelerated rate except for the dark period between 50 B.C. and 1500 A.D. Today, physics is progressing at such a rapid rate that we can only guess what will be discovered and what the consequences will be in the future. Certainly the atomic bomb has indicated how closely the affairs of this world are related to the developments in the physics laboratory, and it can only be hoped that man can learn to get along with his kind as fast as he can discover the secrets of nature, lest things get out of control and he annihilate himself.

In a way this chapter has been something of a summary of the whole field of physics but one developed chronologically rather than logically. Following chapters will deal with much of this material in amplified form but presented in a different order.

QUESTIONS

1. How can the Greek period in physics be briefly characterized?
2. Why were Aristotle's views so widely accepted as late as the fifteenth century?
3. What is meant by the statement that the Arabians left Greek science in "cold storage"?
4. Why is Galileo considered the father of modern science?
5. Why is the classical period often referred to as the Newtonian period in physics?
6. What broad fields characterize modern physics?
7. List several points of impact between science and society.

CHAPTER III

MECHANICAL CONSIDERATIONS

FORCES AND MOTION
VECTOR NATURE OF FORCE, EQUILIBRIUM, LAWS OF MOTION

We now return to the primary purpose of this text, i.e., the logical development of the various concepts of physics. In this chapter the very important concepts of force and motion are to be discussed, because these constitute a logical starting point, and all that follows will be related to them. These two concepts must be considered simultaneously, for neither has physical significance alone. This is because of the fact that *force* is recognized as a *push* or a *pull* which tends to *change* the *motion* of a body. Motion in the abstract is not physical. Only the motion of matter counts, and a force is necessary to produce such a condition. Although motion in the abstract is not physical, a brief consideration of it under the heading of *Kinematics* is desirable for the sake of developing the proper vocabulary for the later study of matter in motion, a branch of physics known as *Dynamics*.

At the same time an opportunity is afforded to develop an appreciation for a very important aspect of many physical quantities, namely the directional aspect. Direction in physics is often fully as important as magnitude, and the realization of this alone tends to discount the purely numerical side of the study, since direction is a geometrical rather than an arithmetical concept.

Vector Nature of Force. The action of a force on a body is a familiar experience. Everyone recognizes the distinction between a large force and a small one, but the fact that force is directional is not always appreciated. Thus, for instance, a force of so many pounds pushing in a northerly direction on a particular body produces no effect whatever in an easterly direction. See fig. 3.1. It affects the motion of the body only in the northerly direction. Moreover, the effect can be completely nullified by a force of the same magnitude pointing in a southerly (opposite) direction. See fig. 3.2. This illustrates the important point that

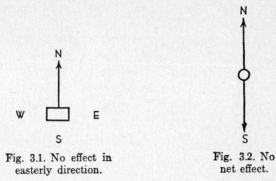

Fig. 3.1. No effect in easterly direction.

Fig. 3.2. No net effect.

when two forces of equal magnitude are added together, i.e., act upon the same body, the result is not necessarily a force with twice the magnitude of one, but one which may even be zero, depending upon the directions involved. The result is a doubled force only when the two forces point in the same direction. See fig. 3.3. When the directions of two forces of *equal magnitude* happen to make an angle of 120° (an angle of 120° is one-third of a whole circle) it can be shown, both by calculation and by experiment, that the resultant force—that is, the force which by itself would produce the same effect as the other two combined—has exactly the same magnitude as each of

Fig. 3.3. Effect doubled.

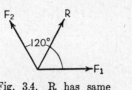

Fig. 3.4. R has same magnitude as F_1 or F_2.

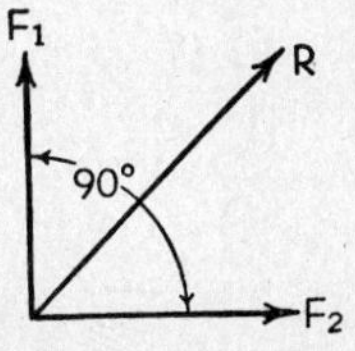

Fig. 3.5. Magnitude of R is approximately 1.4 that of F_1 or F_2.

them (no more, nor less) although of course the resultant direction is quite different from either. See fig. 3.4. Another special case is the one in which two such forces make an angle of 90°—i.e., a right angle—with each other. Here is a force of about 50% greater magnitude than each, which points in a direction lying halfway between the two. See fig. 3.5.

Composition and Resolution of Forces. The student should be interested in knowing how these results are determined. It is a matter of experience, of course, that they are true, but the fact that they can be determined without resort to experiment, except to check the answer, is very important. The simplest way to determine a resultant force is to represent each force graphically by an arrow drawn to scale, so that its length indicates in a relative way the magnitude of the force, and so that it points in the prescribed direction. If, now, a drawing is made with the tail end of the second arrow joined to the head end of the first, each pointing in its proper direction, the straight line drawn from the tail end of the first to the head of the second will indicate by its length and direction (using the same scale) the magnitude and direction of the resultant. See fig. 3.6. This is obviously the diagonal of a parallelogram made by the pairs of the two arrows as sides. Such arrows are called *vectors*, and force is a vector quantity. That is to say, a vector quantity is one which requires the specification of direction as well as of magnitude for its complete determination. Vector quantities, of which there are many in physics, can be added (compounded) by such a graphical method as just described, and of course, the procedure is not limited to just two vectors. There are also rules of mathematical physics by which vectors may be subtracted, multiplied, divided, etc., but it is not necessary for us to consider these here.

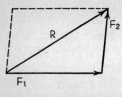

Fig. 3.6. Vector addition.

Thus we see that forces may be compounded to produce resultant forces. A question then naturally arises. May not any force be considered to be such a resultant of perhaps several subsidiary forces—i.e., may not a force have certain components into which it may be broken down for convenience? The answer is obvious, and for practical considerations, very important. A certain force F, pointing in a northeasterly direction, for example (see fig. 3.7), may be considered to be the resultant of a force F_N in a northerly direction combined with a force F_E of the

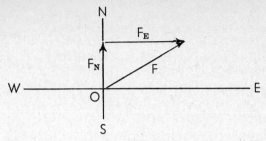

Fig. 3.7. F may be thought of as the resultant of F_N and F_E.

proper magnitude, according to the graphical procedure just outlined, pointing in an easterly direction, even though these two forces may not be apparent. The latter two forces are said to be components of the original one. A force may thus be broken down into any number of components. The projected component of a force in the same direction as the force is the entire force itself, while the projected component of a force at a right angle to itself is zero. See fig. 3.8.

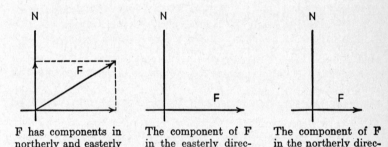

F has components in northerly and easterly directions.

The component of F in the easterly direction is F itself.

The component of F in the northerly direction is zero.

Fig. 3.8.

It should be realized that it is frequently inconvenient to exert a force in the right direction to get its full benefit. For example, it is probably more convenient to tow a barge through a canal by pulling a rope obliquely from the shore than to tow it directly behind another boat. See fig. 3.9. Again, it is probably much more convenient for a tall person to pull a sled by a rope which makes an appreciable angle with the ground, although the full benefit of the pull would be realized only if the pull

Fig. 3.9. Towing a boat by pulling at an angle to the direction of motion of the boat.

were made horizontally. In these cases one is satisfied with only one component of the force exerted. On the other hand, in the case of a sailboat, it becomes possible to sail almost directly into the wind only because the sail can be set at such an angle that the force of the wind perpendicular to it has a component in the direction of the desired motion of the boat. The remainder of the wind's force is dissipated indirectly by means of the keel of the boat. See fig. 3.10.

In these simple cases we have developed the view that not all of a force need be utilized. Lest this create the false impression that the components of a force are always small as compared with the resultant, other cases should be considered. When the angle ϑ between two forces of equal magnitude, acting at a point, exceeds 120° the magnitude of the resultant force is less than that of either force alone. See fig. 3.11. Note that F_1 added to F_2 as previously described yields the resultant R, but also that R is the diagonal of a parallelogram whose sides are F_1 and F_2, and which enclose angle ϑ. As greater and greater angles are considered the corresponding resultants decrease more and more in magnitude. In the limit, when two such forces act in opposite directions (180° apart), as in the example considered early in the chapter, the resultant is zero.

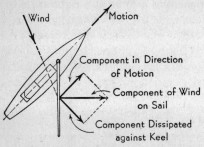

Fig. 3.10. Sailing into the wind. Motion is due to small component of the force on the sail, which itself is a small component of the force of the wind. When the wind blows obliquely upon a sail, the sail appreciates only the component perpendicular to itself; i.e., the wind can push on a sail only perpendicularly.

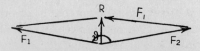

Fig. 3.11. Case of R whose magnitude is much smaller than that of components F_1 or F_2.

Also, when the angle between two forces is almost 180°, a small force acting as the resultant has components along the prescribed directions of much greater magnitude than the force itself.

This is easily tested by experiment. If a heavy package in the shape of a cube is lifted by a cord wrapped about the package,

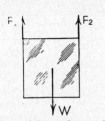
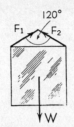
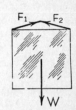

F_1 and F_2 each equal to half the weight of the box.

F_1 and F_2 each equal to the whole weight of the box.

F_1 and F_2 each greater than the weight of the box.

Fig. 3.12.

the cord is quite likely to break when the cord is wrapped quite tightly. On the other hand, if the cord is so loose as to make only a small angle at the point where the cord is lifted, it is not broken so easily. When the cord makes an angle of nearly 180° at the point where it is lifted, as in the case of the tightly wrapped cord,

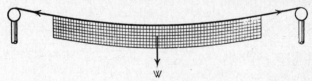

Fig. 3.13. Tension in tennis net approaches infinity as the slack is almost taken up.

the force developed in the direction of the cord becomes infinitely great. See fig. 3.12. This is also why no force whatever is sufficiently great to take absolutely all the slack out of a tennis net if it is applied along the direction of the net. See fig. 3.13.

Equilibrium of Forces. First Condition of Equilibrium. A consideration of the case where the resultant force is zero when a northward component is compounded with a southward component of equal magnitude also clears up another difficulty commonly encountered by the beginning student. It is sometimes hard to understand, if we accept the definition of a force as a push or pull tending to change the motion of a body, how a body can be at rest on a table, i.e., be under the influence of no force, when we know that the force of gravity is always acting. The answer is, of course, that the table also acts on the body, and that the resultant of this force acting upward, compounded with the weight of the body acting downward, is zero, so that the *net* tendency to move is zero. See fig. 3.14. If the table were not there to support it the body would move downward. Thus a body at rest on a table represents a condition called *equilibrium*. It is characterized in part by the fact that the vector sum of all the compo-

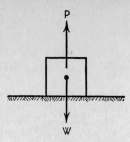

Fig. 3.14. Push of table on body balances pull of gravity on body for equilibrium.

nent forces acting on it is zero. If, for practical purposes, we should restrict our consideration of components to components pointing upward, downward, toward the right, and toward the left, we may say that for equilibrium to exist all components upward must be balanced by all components downward, and all components toward the right must be balanced by all components toward the left. See fig. 3.15. This is what is meant by the statement that the vector sum must be zero. The realization of this makes it possible for the engineer to calculate in advance the forces which will act upon the various members of bridges and other structures when they are subjected to the loads which they experience in use. The basic physical principle behind such calculations is simply this principle of equilibrium.

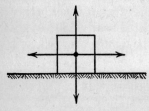

Fig. 3.15. First condition of equilibrium requires that forces up balance forces down, and that forces to right balance forces to left.

This principle is also utilized in the determination of mass by the beam balance, as previously indicated, but here a second condition of equilibrium is also involved. See fig. 3.16.

A little reflection shows that there are two general types of motion into which a body may be set by the action of a force. These are called *translatory motion*, defined as motion in which every straight line in a body remains unchanged in direction throughout the motion, and *rotary motion*, defined as a motion of a body such that every point of the body rotates in a circle about some

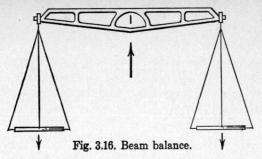

Fig. 3.16. Beam balance.

axis. In the case of the latter, the effectiveness of the force is found to be directly proportional to the perpendicular distance between the axis and the line of action of the force—i.e., it is doubled, tripled, or halved, etc. as this distance is doubled, tripled, or halved, respectively. This distance is known technically as the *lever arm* of the force. See fig. 3.17. If the line of action passes directly through the axis, no rotating tendency is produced because the resulting lever arm is zero. Thus the tendency to rotate

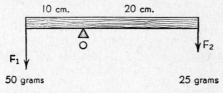

Fig. 3.17. 50 grams at 10 cm. lever arm balances 25 grams at 20 cm. lever arm. F_1 tends to produce counterclockwise motion about the fulcrum O. F_2 tends to produce clockwise motion about the fulcrum O.

depends upon the two quantities, force and lever arm, which give rise to a concept called *torque*. It is defined as the product of a force and its lever arm, with respect to a given axis. Torque is sometimes called *moment of force*. *Clockwise* and *counterclockwise* rotations are thus capable of being produced by the action of clockwise and counterclockwise torques.

Second Condition of Equilibrium. The formal statement of the second condition of equilibrium is that the sum of all torques about any axis must be zero; i.e., all clockwise torques must be balanced by counterclockwise torques. In the use of the beam balance, the pull of gravity on a body (weight of the body) placed in one pan acting with a certain lever arm with respect to the fulcrum is balanced by the pull of gravity on standard masses (weight of the standard masses) placed on the other pan acting with its lever arm. Note! if the lever arm of one pan is doubled, only half the force is necessary to maintain the constancy of the torque, i.e., to produce a certain tendency toward rotation. This is what is meant by the statement that a force is made more effective as regards rotational motion by increasing its lever arm with respect to some axis which serves as a fulcrum. Thus machines such as levers and other mechanical contrivances can be devised to facilitate certain tasks, but a more complete discussion of these matters will be reserved until such concepts as work and energy have been developed.

Need for Study of Motion. In the discussion of forces up to the present point the vector aspect, in particular, has been considered. It is obvious, however, as was noted at the outset, that a complete appreciation of the force concept cannot be had until the vocabulary of motion is developed, for the tendency to change motion is an inherent part of its definition. Motion, on the other hand, is a meaningless concept until the concept of position is established.

Position and Displacement. Position is a relative matter. In the three-dimensional world in which we live, bodies are best located with respect to certain landmarks by laying off specified distances in three mutually perpendicular directions from the landmark taken as an origin. See fig. 3.18. Thus so many paces to the north (or south), then so many to the east (or west), and so many feet up (or down) from a given point will suffice to locate any other point in space such as P with respect to this point O. See fig. 3.18. A change in position involves a displacement from the first to the second position, such as from O to P. This is obviously a directional matter. The concept of change of position in a certain direction is called *displacement*, and is a vector quantity like force. See fig. 3.19. This means that

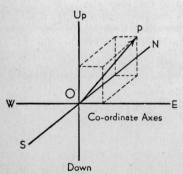

Fig. 3.18. Three-dimensional location of point P (north-south, east-west, up-down).

in technical considerations a distinction must be made between distance and displacement. For example, a car left in a parking lot may later be found ten feet to the right of its original location. It is therefore correct to say that the car has undergone a dis-

Fig. 3.19. Displacement from A to B with respect to the origin O on a straight line.

placement of ten feet to the right, although in so doing it may have been driven a distance of several miles, out of the parking grounds and all over the city, later being returned to the lot and parked ten feet to the right of the original location. Displacement, then, is a technical concept and one which is fundamental in the study of motion.

Velocity. The next concept to be considered is that of *velocity*, the time-rate of displacement, and again the fundamental natures of length and time are brought out. When a displacement takes place in a certain length of time, the ratio of the displacement divided by the time is called the *average velocity* of the body undergoing the displacement. Thus a car traveling twenty miles northerly in one hour is said to have an average velocity of twenty miles per hour northerly. Velocity is thus a vector quantity and is subject to the laws of vector combination. That the direction of a velocity is important is brought out by the example of the rowboat crossing a flowing stream in which the current is appreciable. It is obvious that, to reach a specified point on the opposite shore in a given length of time, the direction in which the rower points his boat is fully as important as the speed with which he rows. *Speed* is defined as the magnitude of velocity, i.e., just the amount of it, not considering direction.

In this consideration the matter of constancy of velocity has not yet entered. Although a car may make a displacement of twenty miles in a given direction in one hour and thus maintain an average velocity of twenty miles per hour, it is common knowledge that the actual rate may be a variable quantity. Thus a distinction must be made between average velocity and *instantaneous velocity*. The latter quantity implies a measurement made over an infinitesimally small interval of time, during which a change would be negligible. In a millionth of a second, for example, it is inconceivable that the velocity of a body can change very much. Therefore the average velocity over such a small time interval is accepted as approximately the actual instantaneous velocity. If this interval is not sufficiently short for a given purpose, then it is agreed that an interval of a billionth of a second should be considered, and so on. It is thus seen that instantaneous velocity is really an abstract concept in spite of the fact that it has become a common everyday term in this automobile age. It is really a limiting value approached by the ratio of displacement to time as the latter becomes infinitesimally small. Strictly speaking, the speedometer of an automobile does not measure velocity but only the magnitude of it, or speed, and the device is correctly named, since it does not indicate direction.

Acceleration. A name is also given to the time-rate of change of velocity. It is *acceleration*. This is a vector concept too, because a change in one direction is not the same as a change in any other direction. Of course there are various ways in which a velocity may change. Either the magnitude or the direction may change, or they may both undergo a change simultaneously. Although the magnitude change—correctly speaking, this is a change of speed—is a common experience, we must not overlook the fact that a change in direction, even with a constant speed, constitutes a perfectly good acceleration. Thus a car traveling in a circular path experiences an acceleration although the speed may not change. The direction of this acceleration is toward the center of the circle.

With this very brief introduction to kinematics the student, if he is disposed toward arithmetic, can already appreciate certain numerical relationships, although as has been stated these are of secondary importance in this study. Thus the product of a constant velocity and the time during which it continues represents the total displacement, a quantity which might not otherwise have been specified in some problem. Similarly, if a velocity changes uniformly, i.e., if motion is uniformly accelerated motion, the foregoing relationship between velocity, time, and displacement holds, providing average velocity is used. In terms of initial and final values the average of any two quantities is one half their sum. Thus a car starting from rest (initial velocity zero) and acquiring a final velocity of 60 miles per hour (88 ft. per second) has an average velocity of 30 miles per hour (44 ft. per second).

In ten seconds it will have been displaced 440 ft. if the acceleration is uniform.

Falling Bodies. We now come to our first consideration of a physical phenomenon involving motion. It is a matter of physical observation that every unsupported body falls in a straight line with constant acceleration, except for the fact that air resistance alters very high speeds. See fig. 3.20. This means that in a vacuum all unsupported bodies have a constant downward acceleration, which has been measured as approximately 32 ft. per second per second (i.e., a gain in velocity of 32 ft. per second each second). In metric units this amounts to 980 cm. per second each second or 9.8 kilometers per second each second. Thus the pull of gravity (weight) alone changes the motion of a body by a constant amount each second, i.e., at a constant rate. This fact has not always been known, and its discovery was a remarkable scientific achievement. The acceleration is not readily measured by direct experiment because a body falls 16 ft. in the first second from rest and 48 ft. in the succeeding second, and a second is a rather short interval of time for precise direct measurements. As has been said before, however, physicists starting with Galileo have devised indirect means of measuring very large and very small quantities, so that today the acceleration of gravity is known with a high degree of accuracy. As we have seen (p. 8), it varies slightly from place to place on the earth's surface. This means also that the weight of a body is not constant but varies as the acceleration of gravity varies. See also p. 40.

The fact that the pull of the earth gives all unsupported bodies the same acceleration was not recognized prior to the time of Galileo. Aristotle had taught that heavy bodies necessarily fall faster than light ones, and even when Galileo demonstrated the fallacy in Aristotle's conclusion by dropping bodies from the top of the leaning tower of Pisa, all were not convinced. Even in Galileo's time it was not altogether clear that the motion which a body acquired was the result of the force acting upon it.

Newton's Laws of Motion. The whole matter of force and motion was finally summarized by Sir Isaac Newton (1642–1727) in the form of three generalizations now known as laws of motion. In the scientific sense they represent a discovery which has stood the test of experiment ever since Newton's time. In simple language they may be expressed as follows:

1) Every body continues at rest or in a state of uniform motion unless a force acts upon it.

2) If a force acts upon a body, the body experiences an acceleration in the direction of the force and proportional in amount to it, as well as inversely proportional to the mass of the body.

3) Associated with every force there is an equal and oppositely directed reaction force.

Significance of the First Law. Let us consider now the significance of these brief but comprehensive statements which have come to be accepted as natural laws. The first one makes it clear that no force whatever is required to keep a body moving at a constant speed in a straight line, but rather that a force is necessary to stop such a motion, or indeed to alter it in any way. In other words it implies that all bodies exhibit a property known as *inertia,* defined simply as that property of the body by virtue of which a force is required to change the motion of the body. The measure of inertia is what is technically known as *mass.* Moreover, a condition of rest is to be thought of as merely a special case of uniform motion for which the velocity is zero. Thus we see that except for the necessity of overcoming friction and air resistance no force at all is required to keep an automobile going at 50 miles per hour over a level, straight road. Perhaps the drivers of "free-wheeling" cars have come to appreciate this point, if only to a limited extent.

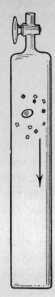

Fig. 3.20. In an exhausted tube a metallic coin and bits of confetti fall at the same rate, showing that air resistance is what ordinarily makes these objects fall at different rates. The acceleration of gravity is the same in a vacuum for all objects at a given place on the earth's surface.

Momentum. The common expression that a car is sometimes carried along by its momentum is a popular version of this important law. The expression is technically correct if the correct definition of *momentum* is understood, as the product of the mass and the velocity of the body concerned. Momentum is another vector quantity and is expressible in terms of arrows. Newton often referred to momentum as quantity of motion, thus emphasizing the view that matter in motion is more meaningful than motion in the abstract. The first law also provides the theoretical basis for the conditions of equilibrium already discussed.

Momentum is a very important concept in physics by virtue of the principle of conservation associated with it. In a system of bodies upon which no unbalanced external force acts, i.e., in which all the forces involved are internal, there can be no change in the total momentum. This is to say that if there is a change in the momentum of one body in the system, this must be offset by a corresponding change in the momentum of other bodies in the system. Thus if two automobiles collide on a slippery road (i.e., no external force exerted by the road on the two cars considered as a system) any change of velocity (magnitude or direction) of the one will be accompanied by a change of velocity of the other, taking into account the masses of each, to keep the total momentum of the system constant. Remember that momentum is a vector quantity.

Another example is the recoiling rifle. Starting with zero momentum, when the trigger is pulled, a bullet of say small mass is projected with a relatively large velocity in a given direction. Treating the bullet and the rifle as a system, for momentum to be conserved, the rifle, of say large mass, must be projected backwards with a correspondingly smaller velocity.

This is also the principle of jet propulsion of airplanes, rockets, etc. and of the maneuverability of space capsules and astronauts by the expulsion of bursts of gases in a direction opposite to that of the desired forward motion.

Significance of the Second Law. The second law explains how motion is changed by the application of a force. In other words it indicates that force is necessary to overcome inertia. A mathematical statement would be that force equals mass multiplied by acceleration. This is why on p. 8 the statement was made that the weight of a body varies slightly from place to place on the earth's surface with the acceleration of gravity. Thus we see that mass is not the same as weight, and that the unit of force (dyne) is equal to the unit of mass (gram) multiplied by the unit of acceleration (cm. per sec. per sec.). This is also why the unit pound should not be used for weight and also for mass at the same time (see p. 8). If weight is expressed in pounds, mass must be expressed in some other unit, and vice versa.

The second law also brings out very clearly that the amount of effort required to change a motion depends proportionally upon the acceleration involved. The reason for powerful motors in automobiles is to provide acceleration ability rather than to enable them to travel fast. What is popularly called "pickup" requires the application of force. This means that if two bodies of the same mass were to be acted upon by different forces, the one acted upon by the larger force would experience the greater acceleration. Furthermore, the same force acting upon different masses would produce a greater acceleration of the less massive of the two bodies. Thus the new fast trains are made less massive than the older ones and require much less time to stop and to come up to speed again at stations by the application of forces presumedly equivalent to those utilized in the more massive trains.

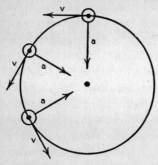

Fig. 3.21. A body traveling in a circular path with constant speed has an acceleration directed toward the center due to a continuous change in the direction of the velocity.

The second law also explains *centripetal force,* the force necessary to keep a body traveling at constant speed in a circular path, and its reaction (see third law below), *centrifugal force,* both of which are involved when cars travel around curves. Unless the friction between wheels and ground is sufficient to provide the suitable centripetal force for a given velocity, the car will skid in a direction perpendicular to the radius of the circular path at the point in question. See fig. 3.21. In this case the acceleration produced by a force is usually the result of change of direction only. The centripetal force acts *on* the revolving body whereas the centrifugal force is exerted *by* the revolving body on something else.

Significance of the Third Law. The third law expresses the observable fact that a force never exists alone. It is impossible to exert a force without a reaction force. By this is meant that for a force to act *on* a body, that body must exert a force of equal magnitude on some other body, i.e., the force *on* the body equals, but opposes, a force *by* the body. A little reflection con-

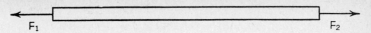

Fig. 3.22. A tug of war is not possible unless F_1 opposes F_2 and vice versa.

vinces one that a one-man tug of war is an impossibility. One cannot exert a force without opposition. See fig. 3.22. The most powerful automobile in the world is helpless in a snow bank where no traction is provided for the wheels, i.e., when the snow is too slippery to develop enough frictional force to push the car out. Actually it is the push of the ground that makes the car start to move. The principle of the conservation of momentum just discussed is a consequence of the first law and the third law of Newton taken together.

Nature of a Scientific Law. It must be remembered that these three laws are accepted today as scientific truths. They represent discoveries in the scientific sense, i.e., according to the methods of science already described. Accepting them as fundamental truths, it is possible to explain all the known phenomena in mechanics. Of course it is possible to set up a logical framework of explanations of these phenomena in terms of other assumed truths, but experience has shown that these assumptions are not only most reasonable and natural—based upon experience —but that they are relatively simple. Thus the physicist today does not attempt to explain Newton's laws of motion. It is sufficient to explain all else in terms of them, and the fact that classical physics can be so explained is one of the marvels of science itself. To be sure, not all of the modern atomic physics has fitted so well into this beautiful pattern, but at least in an approximate way we can say that these laws of Newton's constitute one of the most important landmarks in the science of physics. Today more inclusive theories must be considered, but not to the exclusion of Newton's laws.

Newton's Law of Gravitation. At this point we mention another all-important contribution by Newton, the law of gravitational attraction. In connection with his laws of motion just discussed, it becomes necessary to associate the acceleration of gravity with the force called weight. That is to say, some force of attraction between the earth and an unsupported body is necessary to account for the acceleration which the body has toward the earth. Newton discovered that by postulating universal gravitational attraction between all bodies in the universe he could with the aid of his laws of motion, explain the motions of the heavenly bodies and so concluded that such a postulate was justifiable. This is now an accepted law. It states that every body in the universe attracts every other body with a force that is proportional to the product of their masses and inversely proportional to the square of the distance between them. It is referred to as the inverse square law of gravitational attraction. The proportionality factor, now known as the gravitational constant, was evaluated quite accurately by Lord Cavendish in the latter part of the eighteenth century in England and later by P. R. Heyl of the National Bureau of Standards in Washington, using large, heavy balls and very delicate methods of measuring slight attractive forces.

By this time, the point must be clear that one of the goals of physics is ultimately to be able to explain everything in terms of the simplest assumptions, and express these explanations in the most comprehensive fashion. Thus Newton's laws—three brief statements—and the law of universal gravitation express a wealth of information to those who understand them. Of course it must always be remembered that physics, like all science, never attempts to explain the ultimate "why" of anything, but rather the "how" of it. Another such comprehensive and all-inclusive statement, or law, is the law of the conservation of energy, which we shall consider presently, but first the concepts of work and energy must be developed. The next chapter will deal with these concepts.

QUESTIONS

1. What is meant by a vector quantity?
2. How would the arrow used to represent a south wind differ from one representing a north wind? A west wind?
3. How would an arrow representing an upward pull of 30 lbs. differ from one representing an upward pull of 50 lbs.?
4. If three forces act on an object how would you proceed to find the resultant force?
5. A wire supporting a picture is attached to two points rather close together. How does it happen that if the points to which the wire is attached are rather far apart, the wire might break?

6. If a heavy box is dragged at a uniform rate along a horizontal walk and then up an inclined walk, in which case is the greater force probably required? How is this explained?
7. Does the speedometer on a car register speed or velocity?
8. Can a car traveling at a constant speed have an acceleration?
9. If a spring balance reads 15 lbs. when a given object is placed on it, will it read more or less as it is carried to higher altitudes? Will it read more or less at the north pole? at the equator?
10. Will an equal arm balance which is balanced at the top of a mountain be balanced or unbalanced at sea level?
11. State Newton's three laws of motion.
12. Why does the physicist not attempt to prove Newton's laws of motion?
13. What is the difference between centripetal force and centrifugal force?
14. Distinguish between kinematics and dynamics.
15. What is momentum? Why is it important?

CHAPTER IV

MECHANICAL CONSIDERATIONS (*Continued*)

WORK, ENERGY, AND FRICTION

The Concept of Work. Work is a concept which, in physics, does not have quite the same meaning as in popular usage, where it seems to imply physical exertion. In fact, it is difficult to tell just what is meant by the term "work" in a popular sense, in view of the many arguments over what constitutes and what does not constitute work. It is a technical term here, and as such has to be used in a very exacting manner. When a body is displaced, a force may or may not be acting upon it depending on whether or not the body is accelerated. If a force is acting on a body when the latter is displaced and if the direction of the displacement is the same as the direction of the force, then *work* is said to be done on the body. If the direction of the force should be perpendicular to the direction of the displacement or if there is no displacement, then, technically speaking, there is no work involved. Thus *work* is an abstract concept, defined as the product of force multiplied by the displacement of the point of application of the force in the direction of the force. See fig. 4.1. This is the same as the product of the displacement multiplied by the component of the force in the direction of the displacement.

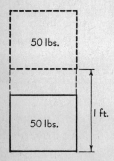

Fig. 4.1. Fifty foot-pounds of work done in lifting fifty pounds one foot vertically upward.

In either event, the two quantities which are multiplied to yield work must point in the same direction, and furthermore, it is necessary (in order for work to be done) that a displacement take place.

From the foregoing it is clear that holding up a wall or a ceiling all day long involves no work at all in the technical sense because no motion takes place, although the process may produce considerable muscular fatigue. Moreover, there would be no work involved at all in "pushing" a heavy trunk at constant speed along a perfectly smooth, frictionless (if such were possible) floor, owing to the fact that, in the absence of friction, the only forces acting on the trunk (i.e., the pull of gravity and the upward push of the floor) act at right angles to the displacement. See fig. 4.2.

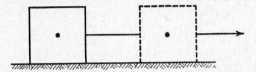

Fig. 4.2. If floor is frictionless, no work is done moving trunk along at constant speed, since no force is required.

This means, of course, that it is impossible to *push* anything (at constant speed) along a frictionless surface because there is no reaction force to push against. Although this frictionless condition can never be wholly attained, it can be approximated as on slippery wax, or ice, etc., under which circumstances the force required is actually very small.

Work, then, is a very special concept, whose importance may not even yet be appreciated. It is measured in foot-pounds or dyne-centimeters or newton-meters. One dyne-centimeter is called an *erg*. It will be recalled that the dyne is a very small unit of force—namely, that required to give a gram of mass an acceleration of one centimeter per second per second. A newton-meter is a *joule*.

Energy—Kinetic and Potential. The capacity to do work is an even more important concept than work. This is called *energy*. Any body capable of doing work is said to be endowed with energy, of which many varieties have been classified. There is energy due to motion, called *kinetic energy*. See fig. 4.3. This is the type of energy possessed by a moving automobile, as indicated by the damage which would result from a head-on collision with another car. It can be shown that kinetic energy is expressi-

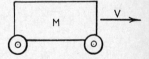

Fig. 4.3. Kinetic energy is possessed by the car if it has a velocity.

ble as one-half the product of the mass and the second power of the velocity.

A boulder on the top of a cliff also has energy because the boulder, if released, is capable of doing work as it crashes down the side of the cliff. See fig. 4.4. This is called *potential energy*. In the grandfather clock, the weights are raised and potential energy is thus given to the clock by the person who winds it. As this energy is released the weights gradually fall, the mechanism is kept in motion, and work is done. Thus potential energy is converted into kinetic energy. Coal, fuel oil, and dynamite, as well as coiled springs and atomic bombs, also possess potential energy because they are all capable of doing work. Thus this abstract capacity to do work becomes a very practical concept of considerable importance.

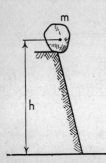

Fig. 4.4. Boulder has potential energy with respect to lower level.

Significance of the Energy Concept. As one contemplates the concept of energy one realizes that it is in all reality a most comprehensive one, for the entire physical world can be thought of as a world of energy. Potential energy stored up in the earth in the form of coal, oil, etc. by the sun—the source of all energy on this planet—is gradually transformed into the kinetic energy of motion, the chemical energy of plants and explosives, the energy of radio, electricity, light, sound, etc. Indeed, each branch of physics may properly be said to be merely a different manifestation of energy, this abstract concept associated with a mathematical product of force and displacement, not to be confused with the popular notions of physical work or vitamin pills.

Conservation of Energy. The concept of energy further provides a possible approach to the whole study of physics because of a very important principle which states that *all the energy in the universe is conserved,* or that this universe can neither gain nor lose any of its capacity to do work. This capacity can be transformed from one kind to another but not lost. The implication is that energy takes on a sort of materialistic significance such that the various branches of physics merely represent different manifestations of it. Nowhere can the store of energy be increased without a corresponding sacrifice elsewhere. It must be remembered, however, that energy really has no such material existence; certainly not in this discussion. It is just a useful concept in terms of which certain explanations become possible, and about which the conservation principle provides the basis for a broad generalization. The generalization is of far-reaching consequences and is on a par with Newton's laws of motion in order of importance. Needless to say, it provides a very useful method of analysis for the theoretical physicist as well as the practical man who designs machinery, for machines do work, even if the amount of work done by a machine is often less important than the rate at which it is done.

Power. The rate at which work is done is called *power*. A large machine can be said to be more powerful than a small one merely because it can do the same amount of work in less time than the small one. It is to be noted that the power of a person, a mechanical device, or an animal should not in any way be confused with the amount of work capable of being done. A race horse, for example, pulling a light buggy at a high rate of speed may do exactly the same amount of work in a given length of time as a draft horse pulling a heavy load rather slowly. The commonly accepted unit of power is the *horsepower*, equal to 550 ft.-lbs. per second or 33,000 ft.-lbs. per minute.

The significance of the power concept is clearly indicated by the present-day methods of paying for labor, i.e., by the hour or by the job. Everyone realizes that for manual labor a man is more valuable than a boy and hence is paid more per hour. On the other hand, a boy working half as fast and receiving half as much pay per hour as the man may accomplish the same total amount of work, and acquire the same total pay, but of course in twice the time. Some jobs, however, call for a faster rate of expenditure of energy than a man can provide, whereupon a more powerful device is required.

Machines. A machine is defined as a device for transmitting and (or) multiplying force. By the use of a machine a large resisting force can often be overcome by the application of a much smaller effort force. In view of the preceding discussion it is now easy to see how this is possible. The amount of work required to be done on the machine is never less than the amount done by the machine—indeed it is always somewhat greater, because of friction—but, since work is the product of a force and a displacement, force can be reduced if the displacement is made correspondingly larger. See fig. 4.5. Thus the force can be halved if the displacement is doubled because the product of the two remains the same. With a block and tackle, for instance, it may be necessary to pull out ten feet of rope to lift a load two feet. See fig. 4.6. Therefore, except for friction, it should be necessary in this case to exert only one-fifth as much force on the rope as would be re-

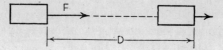

Fig. 4.5. Work here is force F times displacement D. W = F × D. A smaller F acting through a larger D could produce the same answer.

quired to lift the load directly. In this manner *mechanical advantage*, defined as the ratio of the force overcome to the force applied, is acquired. Sometimes a change in direction is the only advantage gained by a machine (see fig. 4.7.), but in all cases the product of the effort force multiplied by the distance through which its point of application is moved equals the product of the

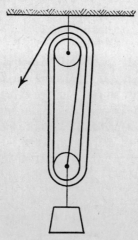

Fig. 4.6. A block and tackle with an advantage of 5 times.

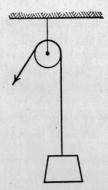

Fig. 4.7. Only an advantage of direction is gained by this pulley.

resistance force (including the resistance of friction) multiplied by its displacement, as a given amount of work is done.

In the absence of friction the mechanical advantage can be

determined merely by considering the relative distances through which the forces are exerted. In fig. 4.6, five feet of rope must be pulled out of the tackle for every foot through which the load is lifted, for if the load is to be lifted one foot, each of the five ropes supporting it must be shortened one foot, but since it is one continuous rope it must be shortened by five feet. This ratio, here equal to the number of strands connected to the movable pulley, is called the *theoretical mechanical advantage* as contrasted with the actual mechanical advantage, which is the ratio of the forces involved. The ratio of the actual to the theoretical advantage is the *efficiency* of the machine. For many purposes only the theoretical advantage is specified since this is readily determined by a consideration of the design of the device. In other words it is usually possible to figure out how far a certain point must be moved by an applied force compared to the distance covered by the resisting force without even operating the machine.

Other Examples of Machines. Another example of a machine is the *ordinary lever*. See fig. 4.8. It is obvious from the above discussion of the pulley that by means of a lever a large force can be overcome by the application of a smaller one because the smaller one acts through a larger distance. The so-called law of the lever, which is readily derived from the law of conservation of energy, states that the force applied, multiplied by its lever arm (previously defined), equals the force overcome, multiplied by its lever arm. With a lever, very little work is required to overcome friction; i.e., the efficiency is relatively high, because the small contact surface between fulcrum and lever does not allow the friction force to operate through a large distance.

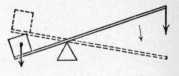

Fig. 4.8. The lever is a machine.

The *inclined plane* also operates upon the same principle as the lever and the pulley. See fig. 4.9. With this device, advantage is taken of the fact that the hypotenuse of a right triangle, i.e., the longest side of the triangle, exceeds the lengths of the

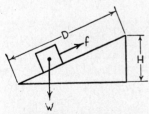

Fig. 4.9. The inclined plane. A small force f acting through a distance D replaces a larger force W acting through a smaller distance H.

other sides according to the size of the angles made between it and them. Thus a heavy load can be raised from the floor to a table by sliding it up a plank stretched from floor to table, making a so-called inclined plane. The length of the incline being greater than the elevation of the table, less force (except for friction) is required to push the body along the direction of the incline than to lift it directly. Even if additional work is required to overcome the added friction introduced by lengthening the path over which the body is moved, the inclined plane often makes possible the raising of loads that would otherwise be impossible because of limitations such as are imposed by human strength.

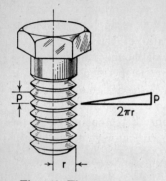

Fig. 4.10. The screw is really a series of inclined planes of circumference $2\pi r$ and pitch P.

A modification of the inclined plane is the *ordinary screw*, a thread of which can be imagined to be unwound so as to produce a wedge-shaped figure which is nothing but an inclined plane. See fig. 4.10. A relatively small force may suffice to produce a complete twist of the screw, causing it to advance, by a distance known technically as the pitch, against considerable opposition. Similarly a relatively weak force applied to the blunt edge of a wedge can produce a large force perpendicular to the face and thus huge logs can be readily split open by forcing wedges into them. The wedge is simply an inclined plane.

In each of the machines listed above (and many more could be mentioned) the fundamental principle is always the same. A small force is applied through a large distance and a large resistance is overcome through a short distance.

Friction. The force of friction is a very important force in physics. Although the engineer is often engaged in attempts to minimize its effects, as by the use of lubricants, to increase the efficiency of a machine, it is nevertheless an important and desirable factor in the daily lives of all. Imagine how one would start even the most powerful or most expensive car upon a frictionless surface.

Coefficient of Friction. The force of friction is characterized by the fact that it always acts along the boundary surface between two bodies. It is thus a tangential force and its magnitude is always proportional to the force pressing the two surfaces together. See fig. 4.11. The pull F due to the weight of the falling body just balances, and is therefore a measure of, the force f of friction when the body is caused to move at constant rate. The proportionality factor is known as the *coefficient of friction*, which is technically defined, for any pair of surfaces, as the ratio

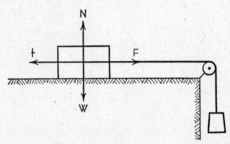

Fig. 4.11. Friction force f is proportional to N; i.e., $\frac{F}{N}$ = constant (coefficient of friction).

of the tangential force of friction divided by the perpendicular force pressing the two together. The force of friction does not, as is commonly thought, depend upon the area of the surface in contact. The coefficient is of course different for different pairs of surfaces.

Summary. In the last two chapters we have seen how force and motion are intimately related. We have considered the vector nature of force, a characteristic which is shared by many quantities in physics. The vocabulary of motion has been developed to the point where it becomes possible to understand Newton's fundamental laws of motion. The concepts of work and energy have been developed and the very fundamental principle of the conservation of energy is thus given meaning. We also have seen how mechanical devices can be used to multiply force, and thus a better understanding of the operation of machines is had. Work, energy, force, and motion comprise that portion of physics commonly called mechanics. Even with this rather brief survey of these topics, limited almost to vocabulary considerations, the student should acquire a better appreciation of this phase of the physical world in which he lives, in addition to a greater respect for the ability of the human mind to put together concepts by which to describe it concisely and unambiguously.

Note: In addition to questions after each chapter, sets of "Review Questions" are introduced between groups of chapters. These are of the "multiple choice" type: each question is presented with several answers, only one of which is correct.

QUESTIONS

1. If a load is to be raised to a point 10 ft. higher than it now is, would you prefer to carry it up a vertical ladder or drag it up an inclined plane? Which requires the greater force? In which case will more work be required? Why?
2. How much work is required to push a 50-lb. body 10 ft. along a horizontal frictionless plane at a constant velocity?
3. What is meant by energy? Why is it important?
4. Why are coal and gasoline said to contain energy?
5. Explain how a force of 20 lbs. can move a 200-lb. body up an incline which rises 1 ft. in 10 ft. if friction is neglected.
6. What is power? Is it possible for a so-called "flea-power" motor and a "horsepower" motor to do the same amount of work?
7. What is meant by the mechanical advantage of a machine?
8. Why is something which is hard to cut placed near the hinge of a pair of scissors?
9. Must a lever always be used so as to gain advantage of force or is a reduction in force sometimes advantageous?

REVIEW QUESTIONS
(See page 217 for answers.)
CHAPTERS I, II, III, AND IV

1. A hypothesis is: 1. a necessary step in inductive reasoning; 2. the same thing as a scientific theory; 3. always successful; 4. never successful; (5) not allowed in physics .... ( )
2. Galileo: 1. was the father of inductive science; 2. always reasoned deductively; 3. was a Greek scientist; 4. taught that heavy bodies fall faster than light ones; 5. lived in the nineteenth century .... ( )
3. Physics is concerned with all of the following except: 1. force; 2. motion; 3. logic; 4. sophistry; 5. relativity .... ( )

4. The kilogram is the basic unit of: 1. mass; 2. length; 3. time; 4. momentum; 5. acceleration .... ( )
5. Measurements in physics are: 1. always precise; 2. usually indirect; 3. necessary to check theoretical speculation; 4. always made with micrometer calipers; 5. never made with micrometer calipers ...... ( )
6. The Greek period in physics is conspicuous because: 1. of Galileo; 2. Aristotle favored inductive reasoning; 3. the Greeks delighted in manual labor; 4. Arabic numbers had replaced Roman numerals; 5. of deductive reasoning .... ( )
7. Physics is primarily: 1. an experimental science; 2. a study in logic; 3. an earth science; 4. applied mathematics; 5. engineering .... ( )
8. Modern physics: 1. completely displaces Newtonian physics; 2. is nothing but Einstein's relativity; 3. started about 1800; 4. explains why light is propagated; 5. is more inclusive than Newtonian physics .. ( )
9. The British standard of length is: 1. the foot; 2. the yard; 3. the inch; 4. the centimeter; 5. the meter .... ( )
10. A force: 1. always keeps a body going in a straight line; 2. can never point upward; 3. is a vector quantity; 4. is necessary to give a body constant motion: 5. is a synonym for power .... ( )
11. The meter is: 1. a unit of inertia; 2. equal to three feet; 3. a fraction of the earth's rotational period; 4. ten times a decimeter; 5. 100 millimeters .... ( )
12. A force of 10 lbs. exerted on a lever 10 ft. from the fulcrum will overcome a load at 2 ft. from the same fulcrum of: 1. 50 lbs.; 2. 50 grams; 3. 200 ft.-lbs.; 4. 200 lbs.; 5. 100 lbs. .... ( )
13. The resultant of two forces of equal magnitude making an angle of 120° with each other is: 1. a force of the same magnitude; 2. a velocity of the same magnitude; 3. a force whose magnitude is equal to the square root of the sum of the squares of the two forces; 4. zero; 5. a force at right angles to either one .... ( )
14. A velocity may change: 1. only in two ways; 2. in direction only; 3. in magnitude and direction; 4. only vertically; 5. only by Newton's law .... ( )
15. The acceleration of gravity is: 1. 32 cm. per sec. per sec.; 2. 32 ft. per sec.; 3. 32 ft. per sec. per sec.; 4. 32 cm. per sec.; 5. zero .... ( )

16. Galileo dropped a large body and a small body simultaneously from the top of the leaning tower of Pisa and showed: 1. that the gravitation theory was right; 2. that there was no ether for light waves; 3. that Aristotle was wrong; 4. that light bodies do have inertia; 5. that confetti always falls as fast as pieces of metal .... ( )
17. Momentum is: 1. mass times force; 2. force times velocity; 3. force times distance; 4. mass times velocity; 5. mass times acceleration .... ( )
18. The Cavendish experiment refers to: 1. the determination of centripetal force; 2. the conservation of momentum in a collision; 3. the gravitational attraction between bodies; 4. the constancy of the stretch of a spring to the force acting on it; 5. the determination of the acceleration of gravity by dropping a ball upon a pan pivoted to make pencil marks on a rotating drum .... ( )
19. A freely falling body falls in the first second: 1. 32 ft.; 2. 32 in.; 3. 980 cm.; 4. 16 ft.; 5. 64 ft. ... ( )
20. To keep a 10-lb. body traveling with constant speed of 5 ft. per sec. in a straight line requires a force of: 1. 50 lbs.; 2. 10 lbs.; 3. 20 lbs.; 4. 2 lbs.; 5. 0 lbs. .... ( )
21. Newton's second law of motion leads to the conclusion that: 1. forces appear as twins; 2. all bodies are attracted toward the center of the earth; 3. force is mass times acceleration; 4. a body at rest continues at rest unless a force acts; 5. weight is the force of gravity .... ( )
22. Mass is the quantitative measure of: (1) inertia; 2. gravity; 3. weight; 4. momentum; 5. displacement .... ( )
23. Velocity has: 1. force; 2. magnitude; 3. displacement; 4. mass; 5. weight .... ( )
24. Galileo lived at about the time of: 1. Aristotle; 2. Democritus; 3. Plato; 4. Newton; 5. Maxwell ... ( )
25. A body traveling in a circle with constant speed: 1. is accelerated; 2. has constant velocity; 3. is not accelerated; 4. does not move; 5. is not affected by gravity .... ( )
26. If body A is projected horizontally at the same instant body B is dropped from the same point

and if air resistance is neglected: 1. B will reach the ground first; 2. A will reach the ground first; 3. A will go farther than B; 4. A will have a smaller displacement than B; 5. the displacement of each is zero .... ( )
27. Acceleration is: 1. the time-rate of displacement; 2. the rate at which distance is covered; 3. the time-rate of change of velocity; 4. the time-rate of change of speed; 5. the time-rate of change of force .... ( )
28. Work is measured in: 1. ft.-lbs.; 2. horsepower; 3. grams; 4. ft.-lbs. per sec.; 5. lbs. .... ( )
29. Work is the product of force and distance: 1. regardless of direction; 2. never; 3. when the two are mutually perpendicular; 4. when the two point in the same direction; 5. only in the absence of friction .... ( )
30. The work done in moving a 10-lb. body 5 ft. horizontally without acceleration on a frictionless surface is: 1. 10 ft.-lbs.; 2. 500 ft.-lbs.; 3. zero; 4. 50 ft.-lbs.; 5. 500 lb. ft. .... (4)
vertically is: 1. 500 ft.-lbs.; 2. zero; 3. 500 joules; 4. 250 ft.-lbs.; 5. 32 ft.-lbs. .... ( )
32. The capacity to do work is called: 1. power; 2. energy; 3. mechanical advantage; 4. momentum; 5. efficiency .... ( )
33. A 2-horsepower machine necessarily: 1. can do twice as much work; 2. works twice as fast; 3. expends twice as much energy; 4. expends four times as much energy; 5. works just as fast as a 1-horsepower machine .... ( )
34. A 138-lb. man climbing a flight of stairs 20 ft. high in 20 seconds displays approximately: 1. .125 H.P.; 2. one-half H.P.; 3. one-quarter H.P.; 4. 138 ft.-lbs. per minute; 5. one H.P. .... ( )
35. If the velocity of a body is doubled: 1. the kinetic energy is quadrupled; 2. the kinetic energy is halved; 3. the potential energy is doubled; 4. the potential energy is halved; 5. the kinetic energy is unchanged .... ( )
36. By the use of an inclined plane to raise a heavy body up onto an elevated bench: 1. more work is probably done; 2. less work is probably done;

3. more force is required; 4. the same work is probably done; 5. the same force is probably required than without the use of the inclined plane .... ( )

37. The mechanical advantage of any machine is: 1. the ratio of the force overcome to the force applied; 2. the same as the efficiency of the device; 3. the ratio of the force applied to the force overcome; 4. always two; 5. always zero .... ( )

CHAPTER V

ELASTIC CONSIDERATIONS

ELASTICITY, VIBRATIONS, AND FLUIDS

The Concept of Elasticity. In the preceding considerations, the concept of force was discussed in its relation to motion, but no direct means of measuring it was suggested except by a spring balance. To understand how this method is possible we must appreciate a fundamental property of matter called elasticity. So far, we have considered only external forces acting upon bodies, and the property of matter known as inertia. Forces also produce internal effects, which of course cannot be completely understood until a study of the structure of matter is undertaken. Although a complete study is not appropriate at this time, a few outstanding characteristics of matter will serve to make clear the point in question. All matter, in addition to exhibiting inertia, is distorted more or less by the application of force and is furthermore characterized by a relative tendency to recover from such distortion, whether it be a change of shape or volume or both. This property is called *elasticity*. It is readily illustrated by the stretched spring. In other words, matter can be thought of as having certain properties. Two of these properties are inertia and elasticity.

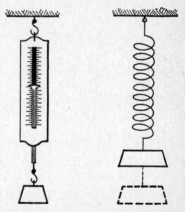

Fig. 5.1. The spring balance operates on the principle of elasticity. The stretch of the spring is proportional to the stretching force.

The Stretched Spring. When a weight is suspended from the end of a vertical spring it is found by experience that the spring is stretched by an amount which is directly proportional to the amount of force acting. This means that the pull of gravity on a body twice as massive as another body will stretch the spring from which it is suspended twice as much as the pull on the lighter body will do. Obviously this furnishes a very simple means of comparing forces. It is the basic principle of the common spring balance. See fig. 5.1. Moreover, it is important to note that when the spring is released it returns to its original unstretched condition unless it has been overloaded, whereupon a permanent distortion is produced and then the elastic limit is said to have been exceeded.

Hooke's Law. The fact that the stretch is directly proportional to the stretching force is a special case of a fundamental law of physics known as *Hooke's Law* in honor of its discoverer, Robert Hooke. In a more general way this law states that when a distortion is produced in a body by the application of force, the restoring force per unit of area (called *stress*), which is developed in all elastic bodies, is proportional to the fractional deformation (called *strain*) so long as the elastic limit is not exceeded. Thus, within the elastic limit, stress is proportional to strain.

Simple Harmonic Motion. In the stretched spring, we have an example of a force whose magnitude varies as the spring is stretched. To put it the other way around, such a force produces a displacement directly proportional to itself. By the application of Newton's second law we note that such a force acting on a given body should produce a motion characterized by the fact that the acceleration also varies directly with the displacement of the body from a position where it is in equilibrium. That this is the case is shown by the following experiment. A spring from which is suspended a massive body assumes a certain stretch. Now if the spring is pulled out farther and then released, it is found that the body bounces up and down. See fig. 5.2. Measurements show that the motion of the body is characterized by just the

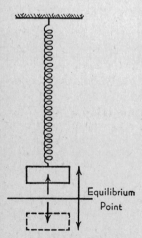

Fig. 5.2. Example of simple harmonic motion

condition noted, namely, that the acceleration at every point is directly proportional to the displacement from the point where it was in equilibrium in a stretched condition. At the equilibrium position the velocity, although having its maximum value, is not changing (a = 0), whereas at either extreme position where the velocity momentarily is zero, the velocity is changing most rapidly (a = maximum). This special motion is called *simple harmonic motion*. See fig. 5.3. The importance of this kind of motion lies in the fact that it is produced whenever a distortion of an elastic body is released and the resultant vibration takes place in a straight line. There are more complicated types of vibrations, of course, but all straight-line vibrations of small amplitude are simple harmonic to a first approximation.

Characteristics of Vibratory Motions. A simple harmonic vibration, being a periodic motion, is often specified by its *frequency*, defined technically as the number of complete vibrations which take place in one second of time. Thus if a boy on a springboard jumps up and down two times per second, his frequency of vibration is two vibrations per second. This is equivalent to saying that one-half second is required for

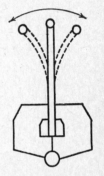

Fig. 5.3. Strip of iron in a vise. Another example of simple harmonic motion.

one vibration. The length of time required for a single vibration is known as the *period* of vibration. The period, then, is the reciprocal of the frequency; i.e., it is unity divided by the frequency. These two concepts, frequency and period, provide the means by which vibratory motions are usually specified in a quantitative manner. The *amplitude* of a vibration is the maximum displacement measured from the equilibrium position, and is not to be confused with the total to-and-fro swing, which is twice the amplitude.

Resonance. All bodies have natural periods, or frequencies, of vibration depending upon their masses, their geometric characteristics, and the manner in which they are set into vibration. Because of this fact the phenomenon of *resonance* is important. A body is found to be set into vibration very readily when a force acts upon it periodically with the body's natural frequency. A child readily learns how to "pump" a swing; i.e., he learns that in a swing he can develop a considerable amplitude if he properly times his impulses in accord with the natural period of the swing. The diver on the springboard also learns how to take advantage of the natural frequency of the diving board to increase his initial velocity.

Resonance effects are sometimes undesirable and are then to be avoided. Frequently it is observed that at certain speeds a car vibrates more readily than at other speeds. This is due to a coincidence of the vibration frequencies produced by irregularities in the road and those produced by the rotating motor and other rotating components of the car such as unbalanced tires, rotating shafts, etc. Two vibrations with the same frequency are said to be in *phase* with each other if both start out together. Resonance occurs when two similar vibrations are in phase with each other and does not occur when they are out of phase.

The Simple Pendulum. Another vibratory motion closely resembling simple harmonic motion is the motion of the simple pendulum. This is simply a small bob suspended by a relatively long flexible cord, and set into to-and-fro motion. See fig. 5.4. The interesting characteristics of pendulum motion were supposedly discovered in the scientific sense by Galileo. He found that the period of vibration of a pendulum depends upon the square root of the ratio of its length and the acceleration of gravity at the place where it is located.

Fig. 5.4. The simple pendulum.

$$T = 2\pi \sqrt{\frac{l}{g}}.$$

He also discovered that the period is the same whatever the amplitude of the swing providing it is not very large. The pendulum of the grandfather clock, although technically not a simple pendulum, because all the mass is not concentrated in a simple bob, illustrates these characteristics. The student can readily infer, and correctly so, that one method of determining the acceleration of gravity at a given place is to measure (rather precisely, of course) the period of a simple pendulum. This method is employed by the United States Coast and Geodetic

Survey, which keeps the scientific world posted with such information at specified locations on the earth's surface.

The Study of Liquids, a Branch of Elasticity. Before going on to a further consideration of vibrations and their effects on matter as a consequence of its elastic properties, a strictly logical presentation of the phenomena of physics requires that we stop here to allow for the study of another group of phenomena which result from considerations of elasticity, namely, liquid phenomena. Elasticity was defined as the ability of matter to recover from distortion. If one were to list various types of distortion, such a list would include simple stretch, linear compression, over-all compression or expansion, and twists or shears. See fig. 5.5. The last of these types, the shear, in which the force is applied tangentially to the surface, plays an important role in physics in that it provides a means of classifying matter.

Matter is readily classified into two main divisions called *solids* and *fluids* according to whether or not resistance is offered to a

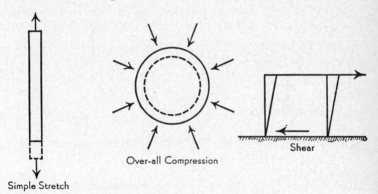

Fig. 5.5. Various types of distortions.

shearing type of force. Although it is all relative and a sharp distinction is not possible in every case, the general classification is useful. The significance of this is that a fluid will not withstand a transverse force. Also, it is impossible to twist a column of fluid. For convenience, fluids are subdivided into two groups, *liquids* and *gases*, depending upon whether or not they display a free surface. Thus a jar can be only half full of water, but with a gas it is different. If the jar contains nothing but air it is always com-

64 ELASTICITY, VIBRATIONS, FLUIDS

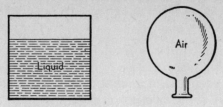

Fig. 5.6. A container partly full of a liquid displays a free surface, but a gas always completely fills its container.

pletely full, whether it contains a little or a lot, because the air displays no free surface. See fig. 5.6. Only its pressure changes. See p. 60 for the definition of fluid pressure.

It is clear that the study of liquids at rest—*hydrostatics*—exemplified by water in free ponds, open columns, wells, test tubes, etc., involves only the action of gravity upon the liquid as a whole, i.e., a simple external force, and does not involve internal forces of the shearing type, because only forces perpendicular to the surface of a liquid can be applied. Thus a free liquid surface is called level because it is perpendicular to the plumb line, which is, strictly speaking, merely the direction of the pull of gravity.

Density. The study of fluids requires special vocabulary. The concept of density, although not restricted to fluid considerations, is all-important to the logical development of this branch of physics because it is one of those concepts in terms of which others are derived. The layman is more or less familiar with the term "density" and, generally speaking, uses it to differentiate between more and less massive substances. For example, a marked

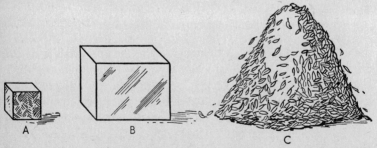

Fig. 5.7. A pound of (A) lead, (B) aluminum, (C) feathers, illustrating differences in density.

distinction between lead and aluminum in terms of mass is readily appreciated, but the technical definition of *density* as mass per unit volume is not such a common notion. See fig. 5.7. Density is defined as the ratio of the mass of a body divided by its volume (grams per cubic centimeter); i.e., it is mass per unit volume. The distinction between weight and mass again comes up. Although density is correctly expressible in pounds per cubic foot (mass-pounds) it is not correct to refer to it as weight per unit volume in spite of loose practice on the part of many technical people. Not that weight per unit volume is an undesirable concept; it simply is not density. Sometimes this is referred to as *weight density*. Again we note that strictly logical thinking often requires painfully precise definitions and careful usage, without which false conclusions may be reached.

Specific Gravity. The term "specific gravity" is also a term heard quite frequently today. Car owners are often concerned with the specific gravity readings of their storage batteries. What does the term mean? Technically the *specific gravity* of a substance is defined as the ratio of the density of the substance divided by the density of water. It provides a way of expressing the relative mass of a certain volume of the substance with respect to the same volume of water. Specific gravity, then, is just a number.

Fluid Pressure. The concept of fluid pressure, already mentioned, will be considered next. Because no shearing force can be exerted on a fluid, it follows that the only force which can be exerted on a fluid must be directed perpendicularly against its surface. Recall the reference to the sailboat on p. 32. Thus the walls of a fluid container push against the fluid, but only perpendicularly. Furthermore, the total force so impressed obviously depends upon the total area upon which it acts. In terms of the perpendicular force per unit area, such a total force is given by multiplying it by the number of units of area. This force per unit area perpendicular to the surface is designated as *fluid pressure*. Thus it makes sense to state that the pressure in a given water main is 80 lbs. per square inch, meaning that upon each and every square inch of surface 80 lbs. is acting regardless of the size of the pipe. An interesting consequence of this is the not so easily realized conclusion that a force of a mere fraction of a pound may suffice to stop a leak if the cross-section area of the leak is only a small fraction of a square inch. Thus a very careful distinction must be made between the concepts of pressure and total force if one is to reason intelligently about these matters.

It can be further shown that pressure at a point below the surface of a given fluid depends upon the depth of the fluid. This of course makes it very easy to calculate the pressure of water at the base of a standpipe, for example. Moreover, since the pressure at a given place thus depends only on the depth, it follows that in two open vessels communicating with each other at the bottom and containing the same liquid, the level of the liquid in one will be the same as in the other. This is because the pressure at the bottom where they communicate with each other cannot have two different values, as would be required if the levels were different. Thus liquids are said to seek their own level under these circumstances. See fig. 5.8. For this reason people often find it necessary to install cellar pumps in houses if the water level in the land about the houses is higher than the basement floors.

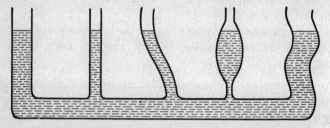

Fig. 5.8. A liquid "seeks its own level" in communicating vessels regardless of the shape of the containers.

Buoyancy. It is a common experience that bodies with specific gravities less in magnitude than unity, i.e., with fractional specific gravities, float rather than sink in water. See fig. 5.9. This brings out an important concept called *buoyancy*, with which the fundamental principle of Archimedes is concerned. When a body is immersed either wholly or partly in a fluid, obviously a certain amount of fluid is displaced. From a consideration of the pressure which the fluid exerts on the body it follows that the net effect of the pressure forces is a resultant force upward which tends to partially offset the downward pull of gravity. This force is called a buoyant force and it can be shown that its mag-

GASES 67

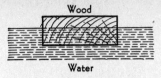

Wood floats in water. Its density is less than that of water.

Iron also floats in mercury.

Fig. 5.9.

nitude is exactly equal to the weight of the fluid displaced. Therefore, if the weight of a body is less than the weight of the fluid which it would displace if submerged, the body will float in the fluid, and vice versa, if it is heavier than the same volume of the fluid it will sink. *Archimedes' Principle* is the formal statement of this thoroughly tested conclusion. It states that all bodies completely or partially immersed in a fluid find themselves buoyed up by a force equal to the weight of the fluid displaced. The principle explains the action of the floating type of hydrometer universally used in garages to determine the specific gravity of automobile batteries. A bob either floats up to a certain mark or it does not, depending upon the specific gravity of the solution in which it is floated. See fig. 5.10. Thus the degree of electrical charge in the battery is determined because it depends upon the specific gravity of the solution.

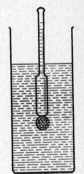

Fig. 5.10. A hydrometer to measure the specific gravity of a liquid.

Gases. According to the definition already given, gases are fluids and so it should, and does, follow that the foregoing applies to the ocean of air which surrounds the earth and constitutes our atmosphere. Air has mass and therefore density. Balloons are buoyed up in air in accordance with Archimedes' principle just as ships are buoyed up in water. See fig. 5.11. By changing the ballast a balloon can be made to rise or fall at will just as a submarine can be made to rise or sink in water.

Atmosphere Pressure and the Barometer. The earth's surface is at the base of the atmosphere and hence a hydrostatic pressure exists due to the weight of the air. This so-called atmospheric pressure amounts to approximately 15 lbs. per square

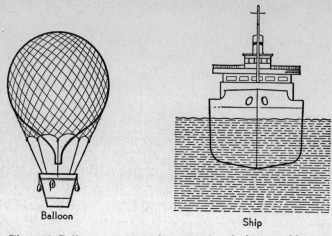

Fig. 5.11. Balloons are buoyed up by the air just as ships are buoyed up by water.

inch and is readily measured by the various types of barometers available today. One can appreciate the magnitude of atmospheric pressure by the following experiment. A slender glass tube some three feet long is filled with liquid mercury. The top is then closed by holding one's thumb over the exposed end. If now the tube is inverted and the end which is closed by the thumb is immersed in a dish of mercury, it is found upon removing the thumb that the mercury does not all run out of the tube. See fig. 5.12. A column some 30 inches high remains in the tube with a vacuum above it. Thus it is seen that the pressure at the bottom, which can be thought of as the place where a tube of mercury communicates with a vessel of air (the whole atmosphere) is just that which exists at a point 30 inches below the surface of mercury, a liquid whose density is between thirteen and fourteen times that of water. Thirty inches of mercury corresponds to approximately thirty-four feet of water, and so lift pumps can raise water up out of a well not more than thirty-four feet. This is because the action of the pump is merely to remove the air from the pipe

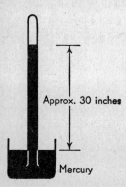

Fig. 5.12. Barometric column.

above the water until the atmospheric pressure forces the water up to take its place.

Measurement of Altitude. The pressure of the atmosphere varies slightly from time to time—almost continuously—due to variations in meteorological conditions. Many predictions of weather are based to a large extent upon barometric pressure. Altitude can also be measured by the barometer, because the pressure decreases with altitude. This is obvious in view of the fact that at high altitudes, the air above a given point weighs less than that above a point at lower altitude, there being less of it in the former case. The aviator utilizes this principle to measure his elevation. The height of Mt. Katahdin, the highest mountain in Maine, was first measured by observing the drop in a mercury barometer as it was carried from the base to the summit of the mountain.

Pascal's Principle. Going back to the discussion of liquid pressure, another interesting consequence follows from fundamental considerations. If two vertical tubes of different cross section are joined together and are partially filled with a liquid so as to leave two open surfaces of different sizes at approximately the same level, then the application of additional pressure to one

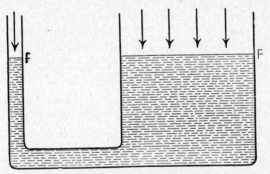

Fig. 5.13. Pascal's Principle. A small force f acting on the liquid in the small tube is able to balance a large force F acting on the liquid in the larger tube in proportion to the respective areas of cross section. Applicable to hydraulic brakes, elevators, barber chairs, etc.

side is transmitted without loss to the other. See fig. 5.13. This is known as *Pascal's Principle*. This principle is utilized in the operation of many practical devices, including the hydraulic brakes on the modern automobile.

As a consequence of the pressure transmission without loss, the total force on the small surface in fig. 5.13 is proportional to the total force on the large surface in just the same ratio as the areas themselves. Thus a 10-lb. force on a 1-inch area is multiplied to 100 lbs. on a 10-inch area because of Pascal's principle, the force per unit area being the same in each case.

Applications of this principle should be obvious. The hydraulic press acts because of its two cylinders of unequal area connected together. In the case of the automobile brake pedal, which creates a certain pressure in a tube of brake fluid, the pressure is transmitted through the pipe line to the brake drums in the wheels, where it acts upon pistons of larger area to create large braking forces. The hydraulic elevator and the hydraulic lift at modern filling stations represent further applications of the same principle.

Summary. In this chapter we have seen that matter displays elasticity. This property accounts for vibratory motions such as are exemplified by springs of all kinds. Simple harmonic motion is the simplest type of vibratory motion. Matter is divided into two classes, solid and fluid, according to its elastic characteristics. Liquids and gases are both fluids. Fluids display several unique properties such as buoyancy, pressure transmissibility, etc. Archimedes' principle gives the relation between buoyancy and fluid displacement. The atmosphere is a fluid which exerts a pressure on all bodies. This is measured by the barometric column, which is also utilized in measuring altitude. Pascal's principle accounts for the operation of many machines such as elevators and hydraulic automobile brakes. Thus it is seen that the study of fluids is a very important part of the study of physics.

QUESTIONS

1. What is meant by elasticity?
2. Would you say rubber is more or less elastic than steel? Why?
3. Describe the operation of the spring balance.
4. Distinguish between stress and strain.
5. What is meant by the period of a vibration?

6. What is the amplitude of vibration?
7. How does the period of vibration of an automobile change when an additional load is placed on it?
8. Describe how the acceleration of gravity might be measured by means of a simple pendulum.
9. What is the technical distinction between a solid and a fluid?
10. Why is water said to seek its own level?
11. How does the "suction" pump really work?
12. Is a "suction" pump suitable for pumping water from a level 37 feet below the surface of the ground? Why? Explain.
13. Is the reading of a barometer affected by carrying the instrument up a mountain providing the temperature is left constant?
14. What is Pascal's principle? Give an illustration.
15. Why does an oil leak in one brake of a hydraulic-brake-equipped automobile ruin the braking action of all four brakes?

CHAPTER VI

WAVES AND SOUND

All the topics thus far discussed under the general heading of elasticity, including the topic of fluids at rest, are commonly grouped under the heading: Statics of Elasticity—a branch of the Mechanics of Deformable Bodies. When deformations are propagated through matter, so-called *elastic waves* are produced, and their study is properly referred to as the study of Dynamics of

Fundamental Wave Concepts. It is a common observation that deformations may be transmitted through material media. A small distortion, such as a localized compression or a transverse shear, readily passes along the whole length of a long coiled spring. See fig. 6.1. Moreover, everyone realizes that a

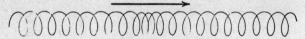

Fig. 6.1. A compression will travel along a coiled spring.

stone dropped into a still pond produces first a depression in the water and then a succession of circular mounds and depressions which spread out and radiate over the entire surface of the pond, moving outward from the central point of disturbance with a definite speed. See fig. 6.2. This phenomenon is called wave

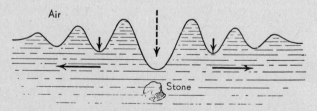

Fig. 6.2. Cross section of the circular mounds and depressions propagated from the point where a stone is dropped into a pond. The medium just vibrates about its equilibrium position.

motion. Technically, *wave motion* is defined as the propagation of deformations through a deformable medium. It is usually a periodic disturbance and is produced by a vibrating source. The medium is set into a succession of vibrations, each specific part moving only back and forth about its own point of equilibrium, but the disturbance travels. It should be noted carefully that the medium itself does not travel. Indeed, it is more accurate to say that only energy travels.

Wave Characteristics. Waves have numerous characteristics. The speed of a wave is a very specific one. By means of a stop watch and a meter stick one can readily time the passage of a given wave across a known distance and therefore determine the velocity of the wave as it radiates out from its origin. The distance between two successive wave crests, or between two successive wave troughs, or more generally between two successive similar configurations, is referred to as the length of the wave or the *wave length*. See fig. 6.3. This is usually designated by the Greek letter lambda (λ).

A traveling train of waves also has a *frequency*, i.e., the number of crests that pass a given point per second. Obviously the following relation holds: the distance between two successive crests (wave length) multiplied by the number of crests per second (frequency) just exactly equals the velocity of the wave

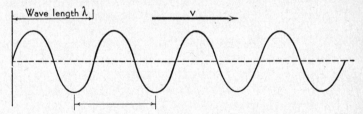

Fig. 6.3. Velocity equals wave length times frequency.

motion. This is probably the most fundamental of all wave relationships. It is also obvious that the frequency of the wave motion is just the same as the frequency of the disturbance producing it. For example, a vibrating reed sets up a wave motion in the medium in which it is located, such as air, with a frequency just equal to the number of times it makes a complete vibration per second. It is also a fact that the length of the wave thus produced depends upon the elastic characteristics of the medium, such as the density and the stiffness, etc. This is usually expressed in terms of the velocity v, which, as has just been shown, depends directly upon the frequency f and the wave length λ.

$$v = f\lambda$$

Another fundamental concept is the *amplitude* of a wave, defined as the maximum displacement from the equilibrium position. See fig. 6.3 again.

Types of Waves. Waves can be classified according to type. There are *transverse waves* such as the ripples just mentioned when a stone is dropped into a still pond. These are characterized by the fact that the displacements of the medium are always perpendicular to the direction of propagation (the surface of the water moves up and down as the wave progresses horizontally). There are also *longitudinal waves* in which the displacements are always back and forth parallel to the direction of propagation. Waves such as sound waves in air are of this type. Compressions and rarefactions of the medium pass along in a direction parallel to themselves. These are the two most common types and the ones to which elementary discussions are usually limited, but torsional, circular, spherical, and ellipsoidal waves, just to mention a few, are considered in more advanced treatises.

Representation of Waves. Whereas fig. 6.3 suggests the appearance of a train of water waves, or in general any kind of transverse wave, it certainly does not illustrate the appearance of

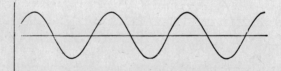

Fig. 6.4. All waves can be represented diagrammatically by the simple wave picture.

a compressional wave. On the other hand, a graph of displacement plotted against time, irrespective of the type of displacement, whether transverse, longitudinal, torsional, or whatever, does produce a curve such as fig. 6.4, which is referred to mathematically as a sine wave. This is because the sine wave is the graphical representation of simple harmonic motion (p. 55) which is, to a first approximation, the type of motion which produces elastic waves. Hence for purposes of representation, a graph such as fig. 6.4 is very convenient.

Further Consideration of the Nature of Waves. Since wave motion has come to be an unusually important concept in physics with the applications of our knowledge of electromagnetic radiations, including light, X rays, and gamma rays, to radio, radar, television, etc., it should be pointed out that the original elastic wave concept has been somewhat extended. As defined above, elastic waves require a medium in which to wave, but our so-called electromagnetic waves appear to travel in vacuum. The representation of waves by the "wavy line" graph (fig. 6.4) suggests a mathematical wave equation since it is merely a graphical representation of a mathematical function of the displacement with time. Today any such observed variation which obeys this so-called wave equation is referred to as a wave motion. These include variations in electric and magnetic fields (defined later) as well as variations in probabilities of electron locations by which atoms and matter generally are described in terms of waves. Thus the wave concept has become much more general than the original elastic wave, but the wave characteristics, properties, and terminology about to be discussed and illustrated by elastic waves apply to waves in general.

Interference of Waves. The wave picture affords, among other things, a simple means of representing the superposition of waves, a common phenomenon. When two waves are superimposed (as for example two water waves) the result depends upon conditions. If the two similar waves of equal amplitude are in phase, i.e., if crest coincides with crest and trough with trough, the result is a reinforced wave of twice the amplitude of either. On the other hand, the result is complete nullification of both if the two waves happen to be just out of phase, i.e., if the crest of one coincides with the trough of the other. There may also be conditions in between complete reinforcement and complete nullification. This general phenomenon is referred to as *interference*, and includes both constructive and destructive types. See fig. 6.5.

Beats. As may be inferred, interference phenomena play a very important role in the study of waves. When two similar

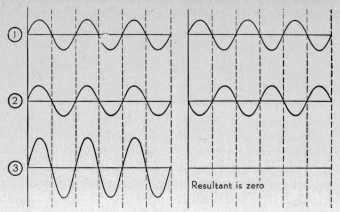

Fig. 6.5. Constructive and destructive interference of waves. The waves in part 3 represent the resultants of those in 1 and 2.

continuous waves of slightly different frequencies are superimposed the result is an alternation of constructive and destructive interference known as the *beat* phenomenon, whose frequency is

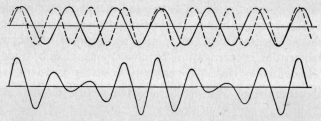

Fig. 6.6. Beats are produced by two waves of almost the same frequency coming into phase with each other every once in a while.

exactly equal to the difference between the two original frequencies. See fig. 6.6. This is easily appreciated in the case of sound waves by striking two tuning forks of nearly, but not exactly, the same frequency.

Reflection of Waves. Waves are *reflected* at boundaries between media. Just as a tennis ball can be bounced up from the ground, i.e., reflected, so can water waves be reflected from an obstruction back into the medium in which they originated. A complete analysis of the phenomenon is not possible here, but suffice it to say that it involves chiefly a further study of the elastic properties of matter.

Refraction of Waves. The changing of speed as a wave passes from one medium to another is technically known as *refraction*. See fig. 6.7. This is another important concept which we shall run across later, particularly in the study of optics.

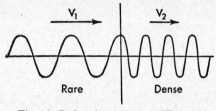

Fig. 6.7. Refraction of waves. Waves undergo a change in velocity—i.e., are refracted—as they pass from one medium to another medium of different elastic characteristics (stiffness and density). Wave lengths are changed accordingly.

Standing Waves. When a wave traveling in one direction in a given medium interferes with a similar wave of the same length traveling with the same speed but in the opposite direction, an interesting situation develops. The result is a wave motion of a very unique type. Every portion of the resulting vibration passes through the equilibrium position at the same instant. There are certain points or regions in the vibrating medium which are always in equilibrium and other points or regions where the disturbance is a maximum. The pattern is called a *standing-wave* pattern because the waves do not progress. They are stationary. The points or regions of minimum disturbance are called *nodal points* or *nodes,* and the points or regions of maximum disturbance are called *antinodes* or *loops.* See fig. 6.8. This condition is realized when a wave passing along a medium meets its own reflection coming back with the same speed and wave length from a boundary. Obviously resonance effects are readily appreciated under these circumstances when the phase relationships are favorable. Owing to the existence of the natural vibration frequencies already mentioned, standing waves are readily set up in bodies. Thus it is natural for a violin string fixed at either end to vibrate in a standing-wave pattern with a loop at the center and with nodes at the ends. Similarly it is easy to

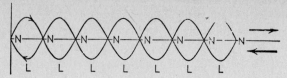

Fig. 6.8. The nodes and loops of standing waves produced by two waves of the same length and amplitude but traveling in opposite directions at the same time.

make a column of air in a pipe, open at each end, assume a standing-wave pattern with loops at either end and a node at the center.

Overtones. As a matter of fact, there are many natural modes of vibration for a string or an air column or for any body at all. Whereas it is natural for a string, such as a violin string, fixed at each end to vibrate with a loop at the center, i.e., as a single segment, it is also natural for it to vibrate in two segments with a third node at the center. Moreover, vibrations in three or more segments are possible. Indeed, it is possible for such a

Fig. 6.9. A string vibrating in (1) one segment, (2) two segments, (3) three segments, (4) four segments—i.e., in its fundamental and 1st, 2nd, and 3rd overtone frequencies.

string or air column to vibrate in any integral number of segments. See fig. 6.9. These various modes are referred to as *overtone vibrations* with respect to the single-segment type called the fundamental mode. The length of the wave of the first overtone in question is obviously just twice the length of the segment. Under certain restrictions one or more of these overtone modes may be impossible. It is often far easier to set a body vibrating in one of its overtone modes than in its **fundamental mode**. It usually requires more energy to do the latter than the former. In general, a vibration is likely to be a combination of the fundamental with several overtones occurring simultaneously. Thus vibrations are often very complicated and may require careful analysis to determine all their components. It is mathematically possible to consider all vibrations in terms of combinations of simple harmonic vibrations. Thus the study of simple vibrations takes on added importance. We shall see presently that all matter is in a state of continuous vibration, because of the vibrations of its component molecules and atoms. From this it may be correctly inferred that vibrations and the waves which are associated with them are ultimately perhaps the most fundamental of all natural phenomena.

Chladni Figures. Vibrations in plates and bars may produce standing-wave patterns of unusual complexity, which can be readily observed by sprinkling fine sand upon them as they vi-

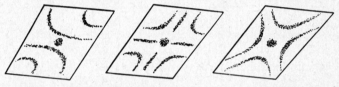

Fig. 6.10. Typical Chladni figures produced by sprinkling fine sand on the plates fastened at their centers and then bowing them with a violin bow, holding them at different points.

brate. The sand piles up at the nodes and disperses in the vicinity of loops. Such designs in sand on vibrating plates are called *Chladni figures*. See fig. 6.10.

Kundt's Apparatus. Standing-wave patterns in horizontal air columns can be observed by the use of fine cork dust strewn uniformly along the bottom of a horizontal transparent tube. When the air is agitated the cork dust piles up in regularly spaced mounds. This device is known as *Kundt's apparatus*. See fig. 6.11. It is used to measure the velocity of waves such as sound waves through a column of air. The distance between the centers of two successive mounds of cork dust is one-half the wave length of the vibration because the nodes are one-half wave length apart.

Fig. 6.11. Kundt's tube for determining the velocity of sound by standing waves. The cork dust piles up in mounds along the glass tube as the rod is vibrated longitudinally.

The wave length multiplied by the known frequency of the vibration gives the desired velocity.

Melde's Apparatus. Standing waves in a cord are readily demonstrated by stretching a cord between a fixed point and the end of an electrically driven tuning fork which sets up the desired vibration. By adjusting the tension in the cord, the latter can be made to vibrate in any one of its natural modes. This apparatus goes by the name of *Melde's apparatus* and is used to determine the precise relationship between the tension, the number of segments, the mass of the cord, etc. See fig. 6.12.

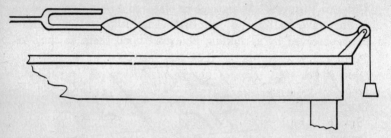

Fig. 6.12. Melde's apparatus consisting of a stretched string fastened to the tip of a vibrating fork. Standing waves are formed along the cord according to the tension set up in it by a weight suspended over a pulley.

Doppler's Principle. One of the outstanding principles associated with wave motion is *Doppler's Principle*. It has to do with the apparent change in frequency of a wave motion as the source of the motion and the observer of it move relatively to each other. It is readily observed by means of sound waves, which will be considered presently. When a source of waves and an observer approach each other, the waves are crowded together so as to produce the effect of an increase in the number passing a given point per second, i.e., an increase in the frequency. See fig. 6.13. Since the velocity depends upon the medium and is therefore constant for a given medium, the length of the wave is correspondingly shortened. The opposite is also true; i.e., the apparent frequency is decreased and the wave length is increased as the source and the observer recede from each other. In the case of sound, the result is a lowering of pitch in the latter case and a rise in pitch in the former. This principle, when applied to light waves, gives the astronomer the remarkable power of determin-

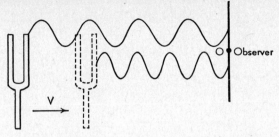

Fig. 6.13. Doppler's Principle. As the fork moves to the right with the velocity V, the waves are crowded and appear to arrive at O with a greater frequency than otherwise.

ing the speeds of certain stars and planets when their motion is directed toward or away from the earth. It is obvious that telescopic means alone are insufficient to supply this information, except when the motion is across the sky, i.e., perpendicular to the line of sight.

Sound as a Wave Phenomenon. Numerous references have been made to acoustical phenomena in this discussion of waves. This is because sound is one of the most common of all the wave phenomena. Sound is nothing but a wave motion in matter (usually air) of a purely mechanical type. Although it is often thought to be restricted to those elastic vibrations capable of stimulating the sense of hearing, there is no particular reason for such a restriction. In a more general sense, the term "sound" applies to any and all elastic waves in matter. Today the field of supersonics is recognized. This is a field in which vibrations above the audible range are considered. The human ear is restricted, in its ability to recognize sounds, to a range of frequencies between approximately 20 and 20,000 vibrations per second. Thus an audible sound is recognized when anything vibrates with a suitable intensity and frequency within these limits in a medium capable of supporting the vibrations and propagating them. Pulses of air suffice to produce sounds regardless of how they are produced, whether by an intermittent blast as through holes in a rotating disk or by the vibration of reeds, bells, air columns, strings, or what not. To prove that sound is an elastic wave motion, let the supporting medium, such as the air, be removed by a vacuum pump, and the sound of a bell, for example, disappears, although the hammer may readily be seen to vibrate if the whole device is mounted in a transparent exhausted glass container. The velocity of sound in air is approximately 1100 feet per second, or 34,000 cm. per sec., or 340 meters per second.

Characteristics of Sounds. Sounds are recognized by three characteristics—loudness, pitch, and quality.

LOUDNESS. Loudness is a function of the intensity of the disturbance. The volume control on a radio governs the loudness of the sounds emitted and, if the set is properly designed, affects no other characteristic.

PITCH. Pitch is the auditory effect of the frequency of sound and the human ear is capable of detecting slight changes in pitch. Pitch is a characteristic by which high notes are distinguished from low notes on a musical instrument. A range of frequencies in which the frequency is doubled is called an *octave*. Thus we have several octaves on the standard piano keyboard such as, for example, the range from middle C approximately (256 vibrations per second) to upper C approximately (512 vibrations per second), which constitutes a single octave. Actually this scale is an old one. On the modern piano this range is from about 261 to 523 vibrations per second. It is the pitch of a note which is affected by relative approach or recession of source and observer according to Doppler's principle already mentioned. It is sometimes difficult to distinguish between the effect of pitch and loudness upon the human ear. The lower limit of loudness capable of being detected by the average ear, or the threshold of hearing, varies considerably with the pitch of the note, a higher intensity being required for the lower frequencies than for the higher ones.

There is a slight discrepancy between the musician's standard of pitch and that of the scientist. Whereas 256 vibrations per second has been recognized as middle C by scientific workers, the musicians favor 440 as A above middle C. This is somewhat different from the 427 which would be required on the scientific scale, and also means that the musician's middle C is 261 vibrations per second instead of 256.

QUALITY. The quality of a sound is determined by the existence of overtones. This means that if it were not for the difference in overtone patterns, a given note would sound just the same when played on the piano, the violin, the cornet, or any other musical instrument. Stripped of their overtones, the same notes played on these several instruments would be indistinguishable, and furthermore such pure notes would lack all semblance of quality. Thus the only reason why a genuine Stradivarius violin may produce more pleasing notes than an imitation instrument is the difference in overtone patterns resulting from the elastic properties of the materials used in construction and from the construction itself. Tuning forks usually produce very pure notes, i.e., vibrations consisting almost entirely of fundamental vibration modes completely free from overtones, and for this reason they are often used in the laboratories for standards of pitch.

Standing Waves in Pipes. Vibrating air columns have the interesting characteristic that when produced in pipes closed at one end and open at the other they lack all the overtones corresponding to the odd integers, considering the fundamental as the zero overtone. Pipes open at both ends are not so restricted in overtone possibilities. See fig. 6.14. Thus it is fair to say that so-called open organ pipes are richer in quality than so-called closed organ pipes. See fig. 6.15.

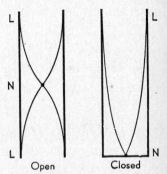

Fig. 6.14. The fundamental wave length of an open pipe is twice the length of the pipe, whereas the fundamental wave length of a pipe closed at one end is four times the length of the pipe.

Analysis of Sounds. The quality of a musical note can readily be determined by the use of instruments which are capable of detecting all the various components in a complicated sound. The trained ear can of course do this to a certain extent, as is evidenced by the manner in which the various instruments can be detected in a symphony orchestra, but electrical apparatus has been developed which can do this more satisfactorily than the average ear.

Summary. We have seen in this chapter that the whole physical side of music is in reality a branch of physics. We have considered the subject of sound as a part of the more general subject of vibrations and their associated waves as fundamental

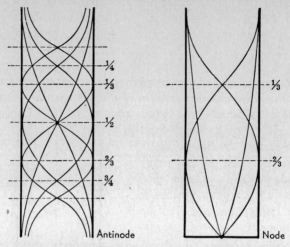

Fig. 6.15. An open pipe (one open at each end) can have twice as many harmonic overtones as a closed pipe (one closed at one end and open at the other). In the closed pipe only odd quarter wave lengths can be fitted into the tube, whereas all quarter wave lengths can be fitted into the open tube, both even and odd—i.e., twice as many.

concepts in this descriptive survey of physics. It has been found that these concepts are just as logically organized and interrelated as the concepts of force, mass, energy, etc., discussed in earlier chapters. Again it has been shown that physics is a highly organized and logically developed science, to understand which the student must acquire a special vocabulary and use it with utmost precision.

QUESTIONS

1. What is meant by sound?
2. Does the frequency of a sound influence the speed at which the sound travels? Do the high-frequency notes of a distant orchestra reach you before the low-frequency notes of the same chords?
3. Is there any part of a gong which remains stationary while it is sounding?
4. Note some of the factors that influence the quality of a sound.
5. Which one of the three factors—pitch, loudness, and quality—enables one to distinguish one orchestral instrument from another?

6. What happens when two sound waves of the same frequency meet out of phase? What is the result if they meet in phase?
7. What happens when two sound waves of slightly different frequencies are combined?
8. As a violinist tightens a string on his instrument is he increasing or decreasing the pitch?
9. Does the loudness of the sound from a tuning fork increase when the fork is set on a table top? Why?
10. Explain the formation of an echo.
11. A fork produces a frequency of 1000 vibrations per second. What could be the frequencies of a second fork which, combined with the first, would give a beat frequency of 5 beats per second?
12. How may sound waves be produced?
13. How is sound energy transferred?
14. After a flash of lightning, thunder may not be heard for several seconds. Why?
15. How does the sound from an open organ pipe differ from that from a closed pipe?
16. A whistle is blown and after 3 seconds an echo is reflected from a cliff. How far away is the cliff?
17. Are sound waves of the order of magnitude of inches, feet, miles, or millimeters in length?
18. The acoustics of a room filled with people are better than those of an empty room. Explain.
19. Can there be too much absorption of sound in a room or auditorium from the point of view of good acoustics?
20. What is meant by the threshold of hearing? Does this threshold vary for different frequencies? The ear is most sensitive to what range of frequencies?
21. Define or state what is meant by "wave motion" and "elasticity."
22. What is meant by the terms "standing wave," "node," "antinode"?
23. What are "beats"?

CHAPTER VII

MATERIAL CONSIDERATIONS

CONSTITUTION OF MATTER, PROPERTIES OF GASES, SURFACE EFFECTS

Our study of physics up to this point has been essentially a consideration of two fundamental properties of matter, namely inertia and elasticity. To go beyond this and acquire an appreciation of additional properties of matter will require a serious consideration of its structure. Although, to some, this appears to be the realm of the chemist, it is well known to the initiated that there is no sharp boundary between the fields of chemistry and physics. This is due in large measure to the fact that the chemist and the physicist working together have, in the last century, made considerable progress in their attempts to unravel the mysteries of matter. Only by the use of the physicist's tools, such as the spectroscope, the X ray, the electron tube, the mass spectrograph, the high-voltage generator to be discussed later, and others too specialized to mention here, have the chemistry and physics of matter advanced to its present stage. Physicists and chemists together have developed theories of the structure of matter around the concepts of molecules, atoms, electrons, protons, and numerous other particles in a continuous state of motion. We shall now consider these concepts in turn.

Molecules and Atoms. Present-day belief is based on the assumption that all bodies are built up out of *molecules*, which are defined as the smallest known units of given chemical substances. It furthermore seems evident that molecules themselves are built up out of smaller component units called *atoms*. In terms of their chemical properties, only some 92 fundamentally different atoms have been recognized, until recently, to occur naturally, and it is felt that all matter is the result of combinations of these elementary units. As recently as the latter part of the last century science accepted the view that the atom was an indivisible entity—hence the name "atom"—but since that time knowledge pertaining to the structure of matter has increased enormously. It seems certain today that the atom has an internal structure of its own. Starting with researches during the very last part of the last century and running all through the present one, the view has become scientifically substantiated that atoms are made up to a large extent of charges of electricity, a concept which will be described presently. A very consistent theory was presented in 1913 by the Danish physicist Bohr, who pictured the atom as a miniature solar system with units of negative electricity revolving about a central, predominantly positively charged nucleus in a more or less elliptical orbit, as the planets revolve about the sun. See fig. 7.1.

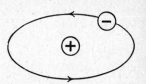

Fig. 7.1. Bohr's conception of the hydrogen atom consisting of a negatively charged electron revolving in an elliptical orbit about a positively charged nucleus (proton).

Although this theory is considered outmoded today, many aspects of the atomic theory of the structure of matter are still popularly described in terms of orbiting electrons, because anything more accurate becomes mathematically too sophisticated to be described except by mathematical equations. But first, what do we mean by positive and negative electricity?

Positive and Negative Charge. The terms positive and negative as used in connection with electricity merely indicate a distinction between two opposite types of electrical charge as determined by the forces which they exert upon other charges,

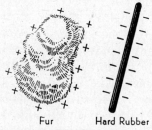

Hard rubber rubbed with fur becomes charged thus.

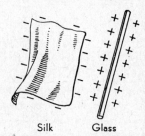

Glass rubbed with silk becomes charged thus.

Fig. 7.2.

either attractive or repulsive. When, for example, a glass rod is rubbed with a piece of silk, or when a piece of hard rubber is rubbed with fur, each of these objects acquires the peculiar property of attracting small bits of paper or other very light objects to itself. This strange force is called an electrical force and is attributed to the existence of something called electrical charge. The glass and the silk act differently, however, and a similar difference is noted between the rubber and the fur, as though one of each pair exhibited an excess and the other a deficiency of this stuff. The distinction was originally indicated by arbitrarily postulating two kinds of electricity called positive and negative just for convenience. See fig. 7.2. But since most electrical manifestations are readily accounted for by supposing the existence of merely a single kind of electricity and considering excesses of it as producing one effect and deficiencies of it as producing the opposite effect, it is not necessary to suppose that two kinds of electricity really exist. The details of these considerations, however, will have to wait for a later chapter on electricity. See pp. 127 ff.

Building Blocks of Matter—Electron, Proton, Neutron, Positron. The smallest unit of so-called negative electricity known to exist is the *electron* (negative). The smallest positive charge is called the *proton*. Bohr's theory suggested that for the simplest known substance, namely hydrogen, the atom consists of simply a proton serving as a nucleus and an electron revolving about it, just as the moon revolves about the earth, in a fixed orbit. See fig. 7.1. The other atoms are similarly constituted but are more complicated, the number of electrons in the various orbits and the constitution of the nucleus varying as the atom increases in complexity from helium, the second substance in the system, to uranium and plutonium, the most complicated. This picture is somewhat out of date already in view of the tremendous rate at which research is being carried on in the universities and other scientific laboratories, but some of its features are still retained by those who require mental pictures for explanations, since the newest theories are mostly mathematical in nature. Recent discoveries include the *neutron,* a neutral particle, and the positively charged electron, the *positron,* so that the building blocks of the atoms are no longer restricted to two in number as was thought for so many years since 1890 or thereabouts. At present it seems quite certain that even the nuclei of atoms have structure, so that whereas only a relatively short time ago the atom was considered the ultimately small and indivisible unit of matter, its structure and the components of its components now concern the physicist and chemist. Such considerations, however, along with the release of atomic energy and the atomic bomb of 1945 must wait for a later chapter.

Brownian Motion. Thus we see that matter is made up of molecules, atoms, electrons, etc.; but more important still for the immediate discussion is the fact that these particles are in a state of incessant motion. We live in a kinetic world, i.e., a world of motion. An indication of this is given by the observation of the so-called *Brownian motion*. This is the continuous agitation which one can see by examining, under the microscope, a sample of water in which fine particles of pulverized carbon are held

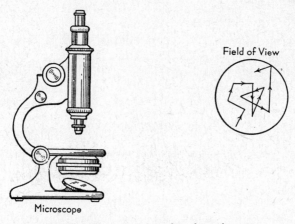

Fig. 7.3. Particles viewed under the microscope are observed to be in a state of agitation. The phenomenon is called Brownian motion.

in suspension because they will not dissolve. Tiny specks of carbon will be seen under these circumstances to be darting about in a most haphazard manner. The phenomenon was first observed by the English botanist Brown, after whom it was named. See fig. 7.3. It is explained by a bombardment of the carbon particles by molecules of the water. As they strike they give rise to the irregular, haphazard motion of the particles which is observable through the microscope. The molecules themselves are much too small to be visible, being only a few hundred millionths of a centimeter in size. They would have to be many thousands of times larger to be visible through even the most powerful microscope available.

Kinetic Theory of Matter. This simple postulate has grown into a theory of sizable proportions commonly known as the *kinetic theory*. To a first approximation it is assumed that molecules are small spherical particles which, by virtue of their internal energy, are bouncing around very rapidly. In the gaseous state the motion is very pronounced and the molecules display a wide range of velocities. Molecules of a given substance display a characteristic average speed. For nitrogen this is approximately 50,000 cm. per second, or nearly one-third of a mile per second, under normal conditions of pressure and temperature. The properties of gases vary noticeably with temperature and pressure.

States of Matter. Matter ordinarily exists in three states: gaseous, liquid, and solid. From the viewpoint of the kinetic theory, the distinction between these states is based upon the proximity of the molecules and the corresponding restrictions placed upon their freedom to move about. They enjoy the most freedom in the gaseous state, but in the solid state they are hemmed in so closely and are acted upon so strongly by intermolecular forces that they are forced to vibrate with very small amplitudes about their equilibrium positions. The transitions from state to state are effected by rather complicated energy transformations. Discussion is usually limited to the gaseous state in elementary courses in physics because this is fundamentally not only the simplest state but the one best understood at the present time. Theory and experiment check very closely for this state. For this reason the theory is often referred to as the kinetic theory of gases.

Boyle's Law and the General Gas Law. One notable success of this theory is its explanation of *Boyle's Law*. This is the law discovered by Robert Boyle in 1662 or thereabouts, which states that a given mass of gas at constant temperature can undergo a decrease in volume only by a corresponding increase in pressure, and vice versa. See fig. 7.4. As a consequence, whatever the pressure or volume of gas, the product of the pressure and the volume of a given amount is always the same. An extension of this to include the effect of temperature is given by *Charles's Law*, and the combined effect of all three factors—pressure, volume, and temperature—is the *General Gas Law*, which states that the volume multiplied by the pressure and divided by the absolute temperature is always constant for a given mass of gas. Absolute temperature will be defined later. Specifically this law means that if one factor is kept constant the variation in the other two is predictable.

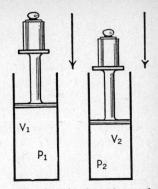

Fig. 7.4. As the volume of a gas is decreased, its pressure is increased, and vice versa if the temperature and mass are left constant.

At constant pressure the volume is increased directly as the temperature is raised. At constant volume the pressure and temperature increase together. At constant temperature, we have Boyle's law itself. Although these results follow from experiment, it is a remarkable fact that they can be derived mathematically from the kinetic theory of gases.

Avogadro's Law and Number. This theory also verifies the postulate put forward by the Italian chemist Avogadro (1776–1856) who followed Boyle (1627–1691), that under the same conditions of temperature and pressure there is always the same number of molecules in a given volume of gas. The number of molecules in a cubic centimeter of air at normal atmospheric pressure and at the temperature of melting ice (these conditions are referred to as normal conditions in scientific work) is enormous. It is the number obtained by moving the decimal point in the number 2.7 nineteen places to the right. This is comparable with the fact that the diameter of a molecule is only 2 or 3 hundred millionths of a centimeter. If one can imagine small spheres packed in a cubical container one centimeter to the side (less than one-half an inch) with each sphere only a few hundred millionths of a centimeter in size, they would be far from crowded. In fact, there would be ever so much more space between the spheres than that occupied by them. The situation has been compared, for simplicity, to a gymnasium approximately 100 ft. by 50 ft. by 15 ft. high containing 50 basketballs. Certainly the hall could not be said to be crowded with basketballs because they could all be swept up into one corner alone of the hall. If, however, they were all in the air at one time, the hall might seem to be full of basketballs, even to the extent that objects on the far side of the hall might be partially obscured by them. This is the approximate manner in which air molecules, for example, are packed into every cubic centimeter of air about us.

Approximate Nature of the Gas Laws. Upon close examination it is found that Boyle's law and the general gas law are only approximate laws. Precise experimentation has shown that there are slight deviations from these laws observable in all cases. Moreover, the deviations become more and more noticeable as we pass from simple gases like hydrogen and helium to gases of greater chemical complexity, and as we pass to extreme conditions of pressure and temperature.

Concept of the Ideal Gas. The scientist has found an ingenious way to cope with this situation. Because all gases obey the gas laws approximately and because the deviation in no case is relatively very large, the concept of the ideal gas has been invented. An *ideal gas* is defined as one which obeys these laws absolutely. Real gases are treated according to their degree of deviation from the ideal gas law. This seems to be a good way of handling the situation because of the complexity which would arise if each case were considered independently. Moreover, the concept has been of inestimable value in organizing the known facts about gases in a general way. Much of the research work today is concerned with the individual details, however, now that it is felt that the main portions of the picture are essentially correct.

The Nature of Ideal Concepts in Science. This procedure exemplifies a characteristic of the modern scientist. Starting with an oversimplified postulate, he continually corrects and revises it as the experimentation progresses, filling in the details as he proceeds. In some cases, strangely enough, the details have not fitted in so smoothly as they were expected to, and often major discoveries regarding main points have thus been made. This has been particularly characteristic of the so-called modern physics. Whereas around 1890 it was thought that all major discoveries in physics had been made and there was nothing left except the job of filling in details by precision-type measurements, quite the opposite has actually been the case. In repeated instances the polishing off, so to speak, has uncovered vast areas of unexplored territory, the developments of which not only have added materially to the store of knowledge, but also have changed entirely the whole point of view. Indeed, we live in an age when people like Einstein, working with the most abstract types of mathematics, have evolved viewpoints of physics in many instances so different from the views held only a few generations earlier, that it is sometimes hard to believe that physics has not become pure philosophy. Many of the concepts of modern physics are purely mathematical in nature and are not capable of translation into physical pictures. Nevertheless, much of the old remains, and it is the belief of some that many of the newer ideas are incapable of being intelligently understood until the shortcomings of the older and more natural views are appreciated. It must never be forgotten, however, that the physics of the practical world, the world of machines and men, is much the same today as ever.

Gas Pressure on the Kinetic Theory. Going back to the kinetic picture again, we readily see how the concept of pressure is explained. Just as a continuous stream of machine-gun bullets will cause a target to act as if it were being subjected to a steady force (see fig. 7.5), so does the continuous bombardment of the walls of gaseous containers by bouncing molecules produce a per-

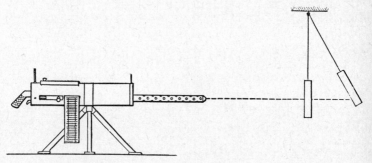

Fig. 7.5. Continuous machine-gun fire develops a steady pressure on a target.

pendicularly directed force which, taken over a unit of area, becomes a pressure in the same sense as the term was originally defined. Moreover, it is clear how this pressure should increase if the volume of a container should be reduced. The molecules would be crowded more closely together so that the frequency, and therefore the force, of their collisions would be increased.

Absolute Zero of Temperature Suggested by the Kinetic Theory. An additional postulate of the kinetic theory is that the temperature of a gas is directly related to the molecular activity, whence it is also clear how the pressure increases with temperature. This suggests a lowest possible temperature in the world as that temperature for which molecular activity ceases. It will be necessary to discuss this point more fully in a later chapter on heat, but suffice it to say here that this is the so-called absolute zero of temperature. It is 459 Fahrenheit degrees below zero, or minus 273° Centigrade. See p. 104.

Effects of Molecular Forces. No discussion of molecular considerations would be complete without mention of some of the effects of molecular forces. One of these is the phenomenon of surface tension. Molecules display forces of attraction between themselves. Forces between like molecules are called *cohesive* forces and those between unlike molecules are called *adhesive* forces. It is the former which hold bodies together and the latter which make different bodies stick to each other. At the surface of a liquid these forces produce an effect suggesting the existence

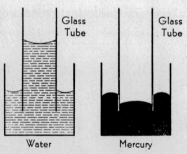

Fig. 7.6. Water sticks to clean glass, but mercury tends to pull away from glass.

Fig. 7.7. A steel needle can be floated on the surface of water.

of a membrane stretched over the liquid so as to make the surface of a given quantity of liquid always as small as possible. The phenomenon is usually referred to as *surface tension*. In certain instances the adhesion forces between two substances exceed the cohesion forces tending to hold the given substance together. For example, water sticks very readily to a chemically clean surface of glass. See fig. 7.6. On the other hand a drop of mercury placed on a plate of clean glass tends to avoid the glass and pulls itself up into a hemispherical droplet. Drops of water on a greasy glass surface do much the same thing.

Capillary Attraction. This same phenomenon also causes water to rise up in small capillary tubes (see fig. 7.6) producing the many so-called capillary effects such as those which keep ink from running out of a fountain pen, keep the moisture in the soil, etc. The opposite effect—i.e., the repulsion of water from greasy surfaces—is utilized in water-proofing garments, tents, etc. It is because of this same effect that insects can walk on the surface of liquids. It is also possible to float a steel needle on the surface of water in spite of the fact that steel is several times denser than water. See fig. 7.7. It is merely necessary that the needle be coated with a very thin film of oil or grease such as it may always have unless specially cleansed by chemical methods. The formation of drops of all kinds is also due to surface tension.

Diffusion. Other examples of molecular phenomena are diffusion and osmosis. If two liquids are brought into contact with each other in a tube, even if care is taken not to mix them together, before long it will be found that one has penetrated into the other. This is called *diffusion*. Gases diffuse very readily, as is evident by the fact that the presence of a volatile substance can be detected by its odor throughout a whole room very soon after it has been released in one corner of it.

Osmosis. Certain substances act like one-way gates for certain fluids. A sugar solution can penetrate in one direction only through the tissues of some fruits and vegetables. Everyone is familiar with the fact that certain fruits when soaked in water take up the water even to the point where they swell up because of the pressure built up by the water. The water does not come out but only goes into the fruit through its surrounding membrane. This phenomenon is called *osmosis*. The pressure developed is referred to as *osmotic pressure*. Osmosis accounts for the rise of moisture up into high trees through their cellular walls to nourish them. The growth of plant life is very dependent upon the existence of osmotic pressure.

Summary. Summarizing the material of this chapter, we have seen that all matter is made up of small units such as molecules, atoms, electrons, etc. We have considered the highly successful kinetic theory of the structure of matter, emphasizing those aspects which apply to the gaseous state. Various other molecular phenomena have also been treated, so that up to this point in our description of physics, we have dealt with the mechanical, elastic, and material phases of the subject. We proceed next to the study of heat and thermometry, which we shall consider from the viewpoint of energy.

QUESTIONS

1. How many molecules are there in a cubic centimeter of air under normal conditions?
2. What is some of the evidence in favor of the molecular theory?
3. A bubble of air becomes larger as it rises in a container of water. Explain.
4. How does one explain the rapid permeation of an odor throughout a room?
5. For what is the physicist Bohr famous?
6. Distinguish between the electron and the proton.
7. What is Boyle's law and what are its limitations?
8. Does a body whose density is appreciably greater than that of water necessarily sink in water?
9. Distinguish between capillarity and osmosis.
10. Explain how rough water is made calmer by pouring oil on it.

REVIEW QUESTIONS
(See page 217 for answers.)
Chapters V, VI, and VII

1. Elasticity is a property of matter by virtue of which: 1. liquids are distinguishable from gases; 2. Archimedes' principle holds; 3. bodies recover from deformations; 4. open pipes have more overtones than closed ones; 5. pressure is dependent upon density .... ( )
2. Simple harmonic motion is characterized by the fact that: 1. the motion is periodic; 2. the acceleration is proportional to the displacement from the

equilibrium position; 3. it is a rotary motion; 4. the displacement is constant; 5. the velocity is constant .. ( )

3. The amplitude of a vibration is: 1. the same as the displacement; 2. the maximum displacement from the zero position; 3. synonymous with the frequency of vibration; 4. the number of vibrations per second; 5. twice the maximum displacement .... ( )

4. Within the elastic limit: 1. stress is always zero; 2. strain is deformation; 3. stress is always proportional to strain; 4. rigidity is negligible; 5. fluids are indistinguishable from solids .... ( )

5. Which of the following names is associated most closely with elastic phenomena: 1. Hooke; 2. Boyle; 3. Archimedes; 4. Newton; 5. Helmholtz .... ( )

6. Ivory soap floats in water because: 1. all matter has mass; 2. all matter has density; 3. the density of Ivory soap is unity; 4. the specific gravity of Ivory soap is greater than that of water; 5. the density of Ivory soap is less than that of water .... ( )

7. When a body is partially or wholly immersed in a liquid: 1. it always sinks; 2. it is buoyed up by a force equal to its own weight; 3. it is buoyed up by a force equal to the weight of the liquid displaced; 4. it is buoyed up because of Archimedes' principle and always floats; 5. it is buoyed up by a force equal to the weight of water which it would displace .... ( )

8. If the mass of a body vibrating at the end of a vertical spring should be increased: 1. the frequency of the vibrations would be increased; 2. the period of the vibrations would be increased; 3. the amplitude would be decreased; (4) the displacement of the body would be decreased; 5. the resonance frequency would be unaltered .... ( )

9. Chladni figures are: 1. wave patterns produced on a vibrating stretched cord; 2. always longitudinal vibrations; 3. always present in sound waves; 4. patterns produced in vibrating plates; 5. the same as antinodes .... ( )

10. When a wave travels from one medium to another of different elastic properties: 1. the wave length is unaltered; 2. the frequency is changed; 3. the velocity is changed; 4. beats are established; 5. sounds are produced .... ( )

11. Kundt's apparatus was demonstrated in the lecture to show: 1. vibrations in plates; 2. vibrations on cords; 3. beats; 4. overtones; 5. standing waves in air columns .... ( )
12. Sound is a wave phenomenon of the following type: 1. longitudinal; 2. transverse; 3. torsional; 4. circular; 5. elliptical .... ( )
13. Standing waves are always: 1. transverse; 2. longitudinal; 3. torsional; 4. the result of interference; 5. the result of refraction .... ( )
14. Kundt's apparatus is used to measure: 1. the velocity of sound; 2. the velocity of any wave motion; 3. the elasticity of a vibrating medium; 4. the number of overtones in a rod; 5. a harmonic series .... ( )
15. Wave length of a wave is: 1. the same as the frequency; 2. the velocity of the wave divided by the frequency; 3. the product of velocity and frequency; 4. always zero in air; 5. the distance from a crest to the next trough .... ( )
16. Whenever two waves of the same frequency, speed, and amplitude traveling in opposite directions are superimposed: 1. destructive interference always results; 2. constructive interference always results; 3. refraction is demonstrated; 4. the phase difference is always zero; 5. standing waves are produced...... ( )
17. In air columns with one end open and the other end closed the wave length of the fundamental mode of vibration is: 1. four times the length of the column; 2. twice the length of the column; 3. always zero; 4. indeterminate; 5. the distance between two nodes .. ( )
18. A node is a point in a wave pattern where the disturbance is: 1. a maximum; 2. a minimum; 3. a rarefaction; 4. a condensation; 5. always varying .. ( )
19. Sound waves: 1. do not travel in a vacuum; 2. travel best in a vacuum; 3. travel with a velocity of 186,000 miles per second; 4. are transverse waves; 5. are always capable of being heard .... ( )
20. Pitch is: 1. the number of vibrations per second; 2. the same as loudness; 3. the same as audibility; 4. the same as intensity; 5. measured in decibels .... ( )
21. Quality of a musical sound is determined by: 1. the loudness; 2. the absence of overtones; 3. the presence of overtones; 4. the pitch; 5. the frequency of the vibration .... ( )

22. The number of beats per second between two tuning forks: 1. measures the frequency of one or the other fork; 2. is independent of the frequency of either fork; 3. is exactly equal to the difference in frequency between them; 4. is approximately equal to the difference in frequency between them; 5. is always audible .... ( )
23. Atoms are: 1. indivisible; 2. invisible; 3. truly spherical; 4. always stationary; 5. composed of molecules .... ( )
24. A long glass tube of mercury with its open end immersed in a dish of mercury can be used to demonstrate the fact that: 1. molecules are small; 2. the atmosphere is made up of nitrogen and oxygen; 3. the atmosphere exerts a pressure; 4. Boyle's law is true; 5. pressure is the result of molecular bombardment .... ( )
25. In order to compress a gas to one-half its initial volume at constant temperature the pressure: 1. must be halved; 2. must be doubled; 3. must be quadrupled; 4. makes no difference; 5. must be atmospheric .... ( )
26. The Brownian movement is: 1. a social reform in physics led by a botanist named Brown; 2. a movement of air currents in the atmosphere; 3. the motion of electrons and protons inside the atom; 4. an optical illusion; 5. a microscopic motion of particles of matter .... ( )
27. On the kinetic theory of the structure of gases, molecules are: 1. large with respect to the distances between them; 2. so small that matter is mostly empty space; 3. smaller than electrons; 4. the same size as electrons; 5. approximately as crowded as a dozen ping-pong balls in a strawberry basket .... ( )
28. Boyle's law has to do with: 1. liquids seeking their own level; 2. the diffusion of gases through porous substances; 3. the law of multiple proportions; 4. Magdeburg hemispheres; 5. the pressure-volume relations of gas .... ( )

CHAPTER VIII

THERMAL CONSIDERATIONS

THE NATURE OF HEAT, THERMOMETRY, EXPANSION, CALORIMETRY, CHANGE OF STATE

In this descriptive study of the physical world in which we live, we have already outlined certain mechanical, elastic, and material considerations and have come to realize the relative importance of such concepts as force, energy, and molecular agitation. The whole complicated picture of the structure of matter is seen to be of the utmost importance to any logical attempt to understand natural phenomena. In considering these phenomena further, questions of a thermal nature arise because many of the effects previously considered seem to be altered if the bodies concerned are heated or cooled. We naturally wonder what is this thing called heat and just what effect it has upon the various properties of matter just discussed. What is really meant by the concepts of heat, cold, temperature, freezing, boiling, radiation, and the many others ordinarily applied, if not always correctly, to thermal phenomena? This is the material to which we now direct our attention in this chapter, in an attempt not only to continue the logical development of fundamental concepts of physics, but also to correct certain popular misconceptions.

Differences between Heat and Temperature. In the first place, the technical distinction between the terms "heat" and "temperature" must be clarified. Heat is something which, if added to a body, usually produces a rise in temperature. Temperature is therefore quite a different concept from heat. Speaking somewhat loosely, temperature may be referred to as a measure of thermal intensity, which, of course, is usually increased as the amount of heat is increased. Temperature is measured in "degrees," whereas heat is measured in units of quantity such as the calorie or the British Thermal Unit, both to be defined presently.

Nature of Heat—Early View vs. Rumford's Interpretation. By the ancients, heat was thought of as a fluid of some kind, the addition of which to a body made it hotter, and the subtraction of which from a body made it colder. This so-called calorific fluid was a concept which seemed to explain many of the observable thermal phenomena, but it had one serious drawback. All attempts to determine the density or any other physical property of the supposedly material substance failed. About 1800 the American-born Benjamin Thomson, later Count Rumford, offered a new interpretation. Engaged in the manufacture of cannon, in which one operation was boring, Rumford observed that although chips were removed there was a noticeable rise in temperature. This was not consistent with the calorific fluid theory because the removal of chips should, on that theory, be accompanied by a loss of heat along with the chips removed. Furthermore, he observed, as others had undoubtedly done before, but he was apparently one of the first to do so critically, that there was more heat generated with a dull tool than with a sharp one, although the latter removed more chips. He sensed the fact that the amount of heat generated was directly related to the amount of energy expended rather than to the number of chips or the amount of material removed. Of course one must work harder with a dull drill than with a sharp one to bore out the same amount of material.

Heat a Form of Energy. Thus was conceived the idea that heat is really nothing more nor less than energy itself. Later experiments by the Englishman Joule and the American Rowland disclosed the direct equivalence of these two. This view is also consistent with the modern kinetic theory of matter, already discussed, in terms of which heat is found to be just that energy associated with the random motion of molecules. On this basis friction develops heat simply because of, and in an amount equal to, the increase in the molecular energy associated with the haphazard, random molecular motion involved when two substances are rubbed together. Indeed, on the kinetic theory, heat is defined simply as the total energy associated with this random, haphazard molecular motion. Moreover, temperature, although qualitatively definable as a property which determines the direction in which heat will flow (it always flows of its own accord from high-temperature regions toward lower-temperature regions), fits into the kinetic picture as the average kinetic energy per molecule associated with the translatory motion.

Thermometry. Before we can pursue farther this matter of heat and temperature it is necessary that we understand how temperature is measured. This phase of the subject is called

Fig. 8.1. The warm water feels hot to the hand previously placed in cold water but cold to the hand previously placed in hot water, showing that temperature is a relative matter.

thermometry. Of course, we might rely upon our sense of warmth and cold, but it can be readily shown that such determinations after it had been in hot water the warm water would feel cold. If, however, the hand is placed in warm water after having been placed in cold water the warm water would feel hot. See fig. 8.1. This shows how relative the matter of temperature is. In order to make a more reliable determination of temperature we look to the various properties of matter which are found to be altered by the application of heat and note first that all matter usually expands when heated. On the kinetic theory, this suggests that the increased molecular activity requires additional space.

Thermal Expansion Utilized in Temperature Measurements. Liquids, generally speaking, expand more noticeably than solids, and gases expand still more. The liquid substance mercury expands very uniformly over a considerable range of temperature and so provides us with a suitable thermometric substance which, in the form of a fine thread in

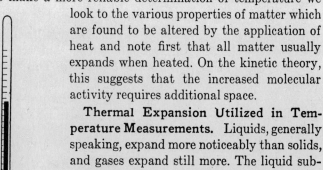

Fig. 8.2. Mercury in glass thermometer.

a glass capillary tube, sealed to an expansion bulb, constitutes a useful thermometer. See fig. 8.2.

All temperature readings are relative. The graduations on a thermometer scale are made to agree with conventional standards, on the basis of which certain temperatures are accepted as fixed points. Two conspicuous fixed points are the temperature of melting ice and the temperature at which water boils under normal atmospheric pressure, i.e., 76 cm. of mercury or 14.7 lbs. per sq. in. It is merely a matter of common agreement that these are the temperatures which are basic to all thermometric work.

Centigrade vs. Fahrenheit Temperature Scales. We have two relatively common scales of temperature: the Centigrade scale, used chiefly in scientific work, and the Fahrenheit scale, which is perhaps more familiar to the layman. On the Centigrade scale the melting point of ice, which is the same as the freezing point of water, is arbitrarily called 0° C., and the boiling point is arbitrarily called 100° C. On the Fahrenheit scale, these are called 32° F., and 212° F., respectively. See fig. 8.3. It is further noted that the Centigrade scale is subdivided into 100 equal degrees between freezing and boiling

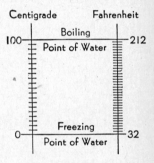

Fig. 8.3. The Centigrade and Fahrenheit temperature scales compared.

temperatures, whereas the Fahrenheit scale is subdivided into 180 equal degrees over the same region, each Fahrenheit degree being 5/9 as large as a Centigrade degree.

Converting from One Scale to the Other. It is obviously a simple matter to change a temperature reading on one scale to the corresponding reading on the other one, since the Centigrade degree is 9/5 as large as the Fahrenheit degree and the freezing point of water on the Fahrenheit scale is already 32 degrees above zero. Thus the familiar temperature 68° F. is 36 Fahrenheit degrees above the freezing point, and so corresponds to only 20 Centigrade degrees above the freezing point, or to just 20° C. Incidentally the temperature reading of 40° below zero happens to be the same on each scale, as can be readily figured out by the student with a minimum of arithmetic.

Coefficient of Thermal Expansion. Whereas different solids and liquids expand quite differently, all gases expand by approximately the same amount for the same change in temperature. The fractional expansion of any portion of a substance per degree of temperature change is defined as the *coefficient of thermal expansion* of the substance. Expansion in one dimension only, i.e., of length, is called linear expansion. We may also consider surface and volume expansion. The *coefficient of volume expansion* for a substance can be shown to be approximately three times the size of the linear coefficient.

Expansion of Gases. The volume thermal coefficient is approximately the same for all gases and is very nearly equal to the the reciprocal of the number 273, per degree Centigrade. This means that, as the temperature of a gas at 0° C. is raised one degree, its volume increases by 1/273rd of its original value. This also means that a gas shrinks in volume 1/273rd of its value if it is cooled one degree Centigrade, and this brings up the interesting question of what would happen to a quantity of gas if its temperature should be reduced by 273 degrees from 0° C.

Concept of Absolute Zero of Temperature. The answer to this question, i.e., whether or not the volume vanishes completely, cannot be found experimentally because of the impossibility of lowering the temperature of any gas to − 273° C. All known gases liquefy before this temperature is reached, and liquids do not have the same coefficients as gases. There is, however, something unique about this temperature of − 273° C. As has been stated before, the pressure of a gas is also related to its temperature, and, curiously enough, the pressure coefficient is such that at − 273° C. the pressure of a gas should also become zero. Now this is significant, for the kinetic theory also requires the pressure of a gas to vanish when the temperature is reduced to absolute zero. Thus, and for more technical reasons not to be considered here, an absolute zero of temperature has been postulated, a temperature so low that, on the kinetic theory, all molecular motion ceases. This temperature therefore is − 273° C., or about − 459° F., and its existence implies the absolute inability of any lower temperature to exist. This is the first place in our study where a limitation has been imposed upon our thinking of extremes of anything such as size or intensity, but it will not be the last one.

Absolute Temperature Scale. Thus a third, or absolute, scale of temperature is suggested, based upon this absolute zero rather than upon the temperature of melting ice. The Centigrade degree is usually retained in this connection with the result that the ice point is + 273° Absolute, and the boiling point is + 373° Absolute. In engineering work, however, an absolute Fahrenheit scale called the Rankine scale is sometimes used.

Further Significance of Absolute Zero. In recent years techniques have been developed by experimenters for producing very low temperatures to learn about these extreme conditions. Although no one has succeeded in producing a temperature significantly lower than absolute zero, and of course, present-day science predicts that no one ever will, nevertheless values within a very small fraction of a degree of absolute zero have been obtained. Incidentally, it should be noted that the measurement of such temperature is by no means a simple matter. The field of low-temperature research appears to hold great promise for startling discoveries in the near future. Many of the properties of matter, particularly of helium, have already been found to vary markedly from their normal values when samples are subjected to very low temperatures. More than one state of liquid helium has been discovered.

Molecular Forces Involved in Thermal Expansion. Going back to the question of linear expansion of solids, attention is called specifically to the magnitude of the molecular forces involved when a rod, say of steel, is heated. Steel increases in length by approximately eleven parts in a million, per degree Centigrade. This means that a steel bridge one mile long, for example, changes in length by more than four feet under a change in temperature from a winter's minimum of 40° below zero to a summer's maximum of around + 100° F. When one considers how much force is required to stretch a steel rod by eleven parts in a million one can only marvel at the quiet manner in which a rise of temperature of only one degree Centigrade produces the same effect. Energy is obviously involved in the process.

Bimetallic Expansion—Thermostats. If a strip of iron is securely fastened along its length to a strip of brass of the same length, and the combination is then heated, the result will be a bending of the combination, because brass expands nearly 50 per cent more than iron. See fig. 8.4. This effect is utilized in the various bimetallic thermostats which are so common in our numerous automatic temperature control devices. In all these devices electric contacts are simply made and broken by such a bending of bimetallic strips upon the application and withdrawal of heat.

Attention is called, incidentally, to the very fortunate fact that cement and steel have practically the same coefficient of expansion. This makes possible the modern steel and cement skyscraper, which does not buckle with changes in temperature. In fact, reinforced concrete in general, which plays such an important part in our highly mechanized lives, is possible only because of this fact.

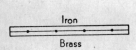 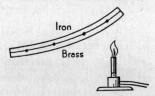

Brass and iron strips welded together, and at room temperature.

The same strips heated. Brass expands some 50 per cent more than iron.

Fig. 8.4.

The General Gas Law. Since the volume of a gas varies with temperature, as indicated above, and since Boyle's law, previously discussed, states how the volume of a given mass of gas depends upon the pressure, it is natural to wonder if these two effects can be combined into a general law. This actually proves to be the case, and for the ideal gas, the relation is a very simple one. It was referred to in the preceding chapter as the *general gas law*. It states that the pressure multiplied by the volume and divided by the absolute temperature of a given mass of gas is constant. This simply means that if the volume of a given mass of gas is kept constant, its pressure varies directly as the absolute temperature. Of course, if the temperature is kept constant this is simply Boyle's law itself, i.e., the pressure is inversely proportional to the volume. And finally, if the pressure is kept constant, the volume varies directly as the temperature. Temperatures, in this connection, must always be expressed on the absolute scale.

Concept of Temperature Only Part of the Study. Thus we see that one of the outstanding characteristics of matter, namely size, is affected by the application of heat, and that variations in the size of an object may be utilized in the measurement of temperature. Of course, this is consistent with the kinetic theory of matter, since an increase in the size of a body can be thought to be associated with an increased molecular activity, which requires more space. For a further insight into the nature of heat we shall be required to consider the question of the quantitative measurement of heat.

Heat Quantities—Calorie and B.T.U. It has already been mentioned that the ancients thought heat to be a material fluid, but that the modern view treats heat merely as a form of energy. For this reason, there would seem to be no need for a new unit in which to measure heat quantity, for the mechanical units of energy or work suffice. This is to say that heat content is readily measurable in foot-pounds, dyne-centimeters, or newton meters (joules). Actually, however, it is more convenient to use other units. Instead of referring to a quantity of heat as so many foot-pounds or dyne-centimeters (ergs) it has become customary to make a comparison with the amount of water that would be heated one degree by this amount of heat. The amount of heat that will raise the temperature of one gram of water one Centigrade degree is defined as a *calorie* of heat. Referring to English units, the amount of heat required to raise one pound of water one Fahrenheit degree is called a *British Thermal Unit* (B.T.U.). One B.T.U. is equivalent to 252 calories. The calorie used by the dietician, however, is the so-called large calorie or one thousand of the above calories. It is clear that these units, namely the calorie and the B.T.U., are defined in a very definite way with respect to a standard substance, water, such that in each case the unit involves a unit of mass and a corresponding unit of temperature in the so-called scientific system and in the popular, or English, system respectively.

Calorimetry and Specific Heat. By thus referring all heat measurements to the rise of temperature of a certain amount of water, a concept called *specific heat* is introduced. It is defined as the ratio of the amount of heat required to raise the temperature of one gram of a given substance one degree Centigrade, to the amount required to raise the temperature of one gram of water one degree Centigrade. A little reflection will show that such a concept is a specific quantity and does not depend upon whether calories or B.T.U.'s are used. It also follows that the amount of heat added to a body to raise its temperature a known number of degrees can be simply expressed quantitatively as the product of the mass of the body by its specific heat, by the rise in temperature. Such a procedure gives rise to the possibility of calculation, if one is so disposed, of heat quantities. For example, the temperature of a hot body of known mass and specific heat may be calculated by immersing the hot body into a known mass of water and recording the temperature rise, upon the assumption that the amount of heat lost by the hot body in cooling is equal to that gained by the water as it warms. This procedure also suggests a method of determining the specific heats of unknown substances, and such calculations are usually referred to as calorimetry. Thus if a 100-gram cube of copper at the temperature of boiling water (100° C.) is dropped into 1000 grams of water at room temperature (25° C.) and the temperature of the water rises to 25.7° C. (while of course the temperature of the copper cube falls to 25.7° C.), then the specific heat of the copper (neglecting the heat absorption of the container) is .094.

States of Matter—Heat of Vaporization. In an earlier chapter the question of the state of matter was raised. It was noted that matter ordinarily occurs in three states: gaseous, liquid, and solid. We now find, as we should expect from the kinetic theory, if heat is energy, that the state of matter depends also upon heat content. A change from one state to another involves the addition or subtraction of a certain amount of heat per gram of substance. Thus when a calorie of heat is added to one gram of water at room temperature and standard atmospheric pressure, the result is a rise in temperature of one Centigrade degree. When, however, the original temperature of the water at this pressure is 100° C., then the result is different. The temperature does not rise a degree for each calorie added, but at this unique temperature, a fraction of the water is changed into vapor. The addition of approximately 540 calories converts the entire gram of water into vapor, and not until this is accomplished does the addition of heat produce a further rise in temperature. During the process the water and the vapor (steam) remain in equilibrium with each other. The amount of heat necessary to convert one gram of water into one gram of steam without changing its temperature is known as the *heat of vaporization* at the temperature in question.

Change in Boiling Temperature with Pressure. It is to be noted also that the temperature at which the liquid vaporizes, i.e., the boiling temperature, is not a fixed temperature but one which depends upon the pressure. The boiling point of water at standard atmospheric pressure is 100° C. or 212° F., but the boiling temperature can be changed. With an increase in pressure the boiling temperature is raised, as is done in the pressure cooker. See fig. 8.5. At high elevations, on the other hand, where the pressure is low, the boiling temperature is also low. In fact it may be too low for certain foods to be cooked by boiling, as is well known by passengers on air liners. Whereas chemistry teaches that substances can be identified by characteristic boiling points, it must be pointed out that it is necessary to establish some standard of pressure such as standard atmospheric pressure, 76 cm. of mercury or 14.7 lbs. per square inch, in this identification process.

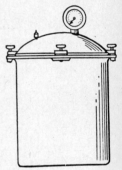

Fig. 8.5. With the pressure cooker the boiling point of water is raised above 100° C. or 212° F.

Boiling under reduced pressure can also be illustrated by the so-called Franklin pulse glass. This is a dumbbell-shaped glass tube, i.e., a slender tube with spherical bulbs at either end, partially filled with colored water and sealed off under reduced pressure. Just the heat of the hand clasped about one bulb causes sufficient rise in temperature to make the water boil over into the other (cooler) bulb.

Boiling vs. Evaporation—Vapor Pressure. In connection with the discussion of boiling it might be well to distinguish between the phenomenon of boiling and that of evaporation. It is well known that water left in an uncovered dish will eventually disappear by evaporation. This is the result of molecular activity at the surface, and is quite understandable from the kinetic theory. The molecules simply bounce away. In the case of boiling, however, the change of state more frequently than not takes place down inside the water, even at the very bottom, where heat is perhaps being applied by a flame. If, however, a cover is placed over the evaporation dish in which there is some water, the molecules are confined above the surface of the liquid, and build up a pressure as they accumulate. At a given temperature, there is a certain maximum value which this pressure can reach for any particular liquid. This is known as the s*aturated vapor pressure*. At this temperature molecules are just as likely to go from the vapor to the liquid as they are to go from the liquid to the vapor. From this point of view the boiling point of a liquid is simply that temperature at which the saturated vapor pressure coincides with the atmospheric pressure.

Humidity. The effects of water vapor in the air afford an interesting study. Everyone knows that atmospheric moisture is a very important factor in connection with the weather. There is obviously a limit to the amount of moisture which the air can hold at a given temperature. If the maximum is reached, precipitation results. This condition is referred to technically as a condition of 100 per cent relative humidity, where the term *relative humidity* is the ratio of the amount of water vapor in a given volume of air to the maximum amount possible at the temperature in question. The actual amount of water vapor present in a given volume of air is called *absolute humidity*. The measurement of humidity is called *hygrometry*. It can be accomplished by the use of the sling hygrometer. See fig. 8.6. Whirling the device keeps the air circulating and enhances the evaporation of moisture from the wet bulb. Because evaporation causes cooling, there is a difference in temperature readings of the two thermometers which is thus related to relative humidity.

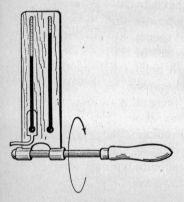

Fig. 8.6. Relative humidity can be measured by the sling hygrometer, in which a wet-bulb and a dry-bulb thermometer are rotated in a frame about an axle. The amount by which the dry-bulb reading exceeds the wet-bulb reading is a measure of the relative humidity.

Dew. Since the limit to the amount of moisture possible in the air varies with the temperature—actually it increases with temperature—we have the phenomenon of the *dew point*. An amount of water vapor which is not sufficient to produce saturation at one temperature may do so, however, at a somewhat lower temperature. Thus the ice-water pitcher often displays condensed moisture, or dew, all over its surface on so-called humid days, because the temperature of the pitcher is sufficiently lower than that of the surrounding air for the amount of water vapor present to produce saturation. Similarly, dew is formed on the grass on those evenings when the temperature is lowered sufficiently, probably by radiation, for the amount of water vapor present to represent saturation. At lower temperatures the vapor instead of condensing changes directly to a solid state called frost.

Freezing—Heat of Fusion. This brings up the other important change of state, the change from liquid to solid and vice versa, called freezing and melting respectively. We have seen that the addition of heat to a substance raises its temperature—for water it amounts to one Centigrade degree per gram—except at the boiling temperature where the change of state is effected. In a similar but opposite way the temperature 0° C. is a unique temperature for water. The subtraction of heat at this temperature produces the change from the liquid to the solid state. For water, 80 calories per gram are required at 0° C. Thus in order to freeze water at this temperature 80 calories per gram must be subtracted, and the melting of ice at this temperature absorbs 80 calories per gram. The amount of heat necessary to change the state of one gram of a substance from liquid to solid, or vice versa, without changing its temperature is called the *heat of fusion* of the substance. For water it is 80 calories per gram.

Refrigeration—Utilization of Heat of Fusion. It is this heat of fusion which is so important in ice refrigeration, which not too long ago was the common method of household refrigeration. Food can be kept cold in an icebox by the melting of ice. Conditions have to be such as to insure that the 80 calories per gram come from the food and not from the outside. Hence food and ice are surrounded by an insulated box, and the efficiency of the icebox is directly proportional to its heat insulation.

House Heating—Utilization of Heat of Vaporization. On the other hand it is the heat of vaporization that is so important in steam heating systems. The 540 or so calories liberated by every gram of steam which condenses in the radiators contribute largely to the heating of the radiators, which in turn heat the air in the rooms of a house in which this system is used.

Sublimation. Many substances are capable of passing directly from the solid to the gaseous state without going through the liquid state. This phenomenon is known as *sublimation*. Everyone is familiar with the fact that so-called dry ice, i.e., solid carbon dioxide, simply evaporates without melting at atmospheric pressure. This is sublimation. Much of the snow on the ground in the winter sublimes rather than melts.

Liquid Air. There was a time when the liquefaction of many gases was impossible, and it was thought that vapors could thus be distinguished from gases, but today no such distinctions are attempted because it is known that all gases can now be liquefied and even solidified. Liquid air, for example, is a very familiar substance in the scientific laboratory. It, of course, boils freely at room temperature and atmospheric pressure, although its temperature is 180° below zero Centigrade. One has to be careful not to confine liquid air in a stoppered container because at room temperature its saturated vapor pressure is many, many atmospheres. For the same reason it is not safe to confine dry ice in a stoppered container at room temperature. The vapor pressure which it will build up is enormous.

Effect of Pressure upon the Freezing Point. Thus it is seen that the state in which matter exists depends among other things upon its thermal condition. At the boiling point the liquid and gaseous states exist in equilibrium with each other, and at the freezing point the liquid and solid states are in equilibrium. Although the boiling temperature of water is appreciably altered by a moderate change in pressure, the freezing point is only slightly changed. Enormous pressures, however, can reduce the melting point appreciably. Ice-skating is in reality a process of slipping along on a thin film of water immediately beneath the skate's runner, a condition produced by local melting of the ice under the pressure exerted by the weight of the skater acting upon the relatively small area of surface supporting the skate. Upon the removal of the pressure, after the skater slips along, this thin film of water freezes almost instantly. *Regelation* is the technical name for this phenomenon of melting under pressure and refreezing upon its removal. See fig. 8.7. It also explains the manner in which the glaciers probably slipped along over large sections of the earth at an earlier period.

Fig. 8.7. Regelation illustrated by a wire which gradually "cuts" through a piece of ice owing to weights fastened at its ends. The ice fuses together after the wire passes through successive portions.

The Triple Point. The lowering of the boiling point by reducing the pressure on a liquid, combined with the change of the freezing point by changing the pressure, raises the question whether or not a proper combination of pressure and temperature can be found for which the liquid both boils and freezes at the same time. This in fact is the case. At a pressure of only a fraction of a centimeter of mercury (4.06 cm. hg) and a temperature slightly higher than 0° C. (.0075°), water boils and freezes simultaneously. This combination is referred to as the triple point. Because this triple point temperature for water is so specific, it has recently been chosen as a better fixed thermometer point than the temperature of freezing water at standard atmospheric pressure (76.0 cm. of mercury) for the calibration of thermometers. The Centigrade temperature scale for which this is the zero point is referred to as the Celsius scale.

Summary. In this chapter we have seen how heat is simply a manifestation of energy, how it is distinguished from temperature, and how the latter is measured by virtue of certain thermal properties of matter, particularly expansion. We have also considered briefly how heat is measured and how thermal considerations enter into questions involving states of matter, and change of state. We shall turn next to a descriptive consideration of heat transfer and allied phenomena.

QUESTIONS

1. Why does the mercury thread in a glass thermometer usually fall a little when the thermometer is first placed in hot water?

2. A pyrex dish is less likely to break when heated than an ordinary glass dish. What can be said of the relative expansion coefficients of these two substances?
3. Is there any sense to the procedure of dipping a metal can cover in hot water to make it come off more easily?
4. What is the distinction between heat and temperature?
5. Is it technically correct for the weatherman to state that temperatures may be expected to be warmer or cooler as the case may be?
6. A Centigrade thermometer reads $-40°$. What would be the reading on a Fahrenheit thermometer?
7. It is often possible by gentle heating to remove the stopper from a glass bottle after it has become stuck. Explain.
8. Is a gallon of boiling water any hotter than a pint of it? Does the one contain more or less heat than the other?
9. Why does steam at $212°$ F. cause a more severe burn than boiling water at $212°$ F.?
10. Why does a gas become heated on compression and cooled on expansion?
11. Explain why the air escaping from an automobile tire valve feels cold.
12. Explain why it is impossible to skate on glass.
13. The substance hypo is often used instead of ice on indoor skating rinks. What characteristics does it have to make it suitable for this purpose?
14. Why does it become necessary to add moisture to the air of a house in the wintertime?
15. Explain why moisture collects on the outside of the ice-water pitcher.
16. Explain why perspiration is more bothersome on a humid day than on a day less humid even if the temperature is the same.

CHAPTER IX

THERMAL CONSIDERATIONS (*Continued*)

THE NATURE OF HEAT TRANSFER, QUANTUM THEORY, AND CERTAIN OTHER PHILOSOPHICAL CONSIDERATIONS

Heat Is Transferable from Place to Place. In the last chapter it was noted that temperature could be thought of as a property which determines the direction of heat flow from one body to another body in contact with it. That heat does flow from one region to another is not an unfamiliar fact, and so a study of the manner in which it flows becomes an important part of any course in general physics.

Heat (Not Cold) Flows. In the first place, it should be clearly understood that heat always flows of its own accord from regions of high temperature to regions of low temperature just as water flows of its own accord downhill. This means that if heat does flow from low-temperature regions to high-temperature regions it must be forced—i.e., work has to be done on it. Furthermore, this means that "cold" does not move. Cold is the absence of heat. It is only the latter, the energy associated with the random motion of molecules, that flows. Thus it is not sensible to talk about keeping the cold out of a tight house on a winter's day. This is not to deny that cold *air* can be forced into a house on a windy day through spaces between ill-fitting windows, doors, etc. In reality the problem is to insulate the house so as to keep whatever heat there is inside from escaping to the outside. Similarly we wear so-called warm clothes in the winter to keep the body heat in and not to keep the cold out. Thus the practical matter of heat insulation is nothing more nor less than the prevention of heat flow and is therefore a part of the larger question of heat flow, which we shall now consider.

Three Modes of Heat Transfer (Conduction, Convection, Radiation). There are three commonly recognized ways in which heat may be transferred from one region to another. These are technically known as *conduction, convection,* and *radiation.* Our study will deal with each of these in turn.

Conduction. When an iron poker is held with one end only in a fire, heat is observed to travel from the hot end to the cold end, passing successively from point to point throughout the length of the poker. This is the type of transfer called *conduction.* In terms of the kinetic theory the molecular agitation is simply passed along from molecule to molecule in much the same manner as a

Fig. 9.1. If the end domino in a row is pushed over, kinetic energy will be passed on from domino to domino. This is analogous to the process of heat conduction.

column of closely spaced, standing dominoes is toppled over if the first one in line is pushed forward. Each domino in succession is pushed over as energy is transmitted through the system. See fig. 9.1.

CONDUCTIVITY. Different substances have different *coefficients of conductivity,* determined by the rate at which heat is conducted through unit length of the substance when a unit difference in temperature is maintained between the ends of the sample of unit cross-sectional area. The natural element which is the best known conductor of heat is silver, which, incidentally, is also the best known conductor of electricity. Insulators are simply poor conductors. Whereas the metals are good thermal conductors, substances like cork and wool are good insulators. Air also like most gases is a poor conductor, i.e., a good insulator, and much of the merit of a fur-lined coat lies in the poor conductivity of the air spaces between the hairs of the fur. This is the same reason for double-walled construction of houses in which an air space is left between the walls. Because powdered asbestos is also an exceptionally good insulator, i.e., a poor conductor, it is commonly blown into the space between the walls of old houses these days to improve the insulation. It should be noted, of course, that if heat could be prevented completely from escaping through walls, windows, roofs, etc. houses could be heated at practically no expense at all. This is why so much attention is given to the question of insulation in "the house of the future."

The advent of the electric refrigerator focused considerable attention upon the matter of heat insulation. For reasons of operat-

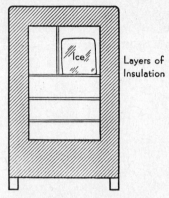

Fig. 9.2. Insulation is the secret of the successful icebox. Heat is prevented from flowing in from the outside to melt the ice. Only heat from the food inside the box is allowed to melt the ice.

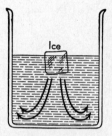

Fig. 9.3. Convection currents are set up by an ice cube floating in a glass of water. Cold water sinks and gives way to warm water which rises.

ing economy the early builders of electric refrigerators were forced to develop boxes insulated to such a degree of refinement that the manufacturers of iceboxes learned how to build boxes that needed to be filled with ice only once a week. It is entirely a matter of preventing the flow of heat from the outside to the inside of the box that accomplishes this result. See fig. 9.2.

Convection. The second mode of heat transfer, which we shall now consider, is convection. This is characterized by the word "circulation." If an ice cube is placed in a glass of water it is found that currents are established in the water. See fig. 9.3. The coolest water sinks to the bottom of the glass and the warmer portions rise to the top. This is due to the difference in density between warm and cold water and the natural application of Archimedes' principle of buoyancy already described. Such currents are known as convection currents, and the phenomenon of heat transfer by such means is called *convection*. It is characterized specifically by the acquisition of heat by some material agent which moves along and carries heat with it from one place to another.

The heating of the rooms of a house is accomplished in large measure by convection currents of air rising from heated radiators or registers. It is to be noted that those houses which heat best are those in which a good circulation of air is provided by the arrangement of the rooms. The old-fashioned long, narrow house with rooms following one after another is hard to heat uniformly. This is not the case with houses in which the rooms are clustered

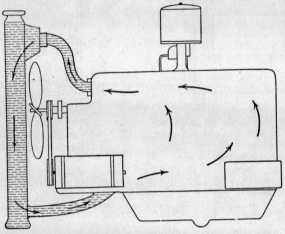

Fig. 9.4. Cooling is accomplished in the radiators of certain cars by convection currents. Hot water in the engine rises and is allowed to fall through tubes in the radiator, thus becoming cooler and displacing warmer water at the bottom.

around some central point with open doorways providing free access from one to another, including the last and the first.

The principle of heat transfer by convection is still utilized in the cooling systems of some automobiles, although most of the cars today have water pumps for this purpose. The hot water in the engines of these cars rises to the top of the radiator where it falls through narrow tubes of brass from which, by conduction, much of the heat is removed to the out-of-doors. At the bottom of the radiator the water becomes cool, forcing more hot water up through the motor to the top of the system, where the process is repeated. This is the so-called thermosyphon system. See fig. 9.4.

METEOROLOGICAL ASPECTS OF CONVECTION. Convection currents of air play a very important role in the production of weather. Heated air masses rising rapidly by convection on a hot day are the cause of most thundershowers. The electrical phenomena involved in such showers will be discussed in a later chapter, but the cloud formations responsible for them are produced by convection currents in which heat is carried high up into the air from heated areas on the earth's surface. Studies of air masses at very high altitudes, and subject to convection currents, have played a very significant role in modern meteorology.

Radiation. The last of the three modes of heat transfer to be considered here is radiation. It is a known fact that a person's hand placed a few inches beneath a hot household radiator feels warmth from the radiator. See fig. 9.5. Similarly, warmth is felt a few inches from the floor in front of a glowing fireplace fire. In neither of these cases is the transfer of heat due chiefly to conduction or convection. In the first place air is a very poor conductor, and in the second place convection carries heat upward and not downward. In these cases the heat is *radiated* directly outward; i.e., it is transferred much as light is transferred by some kind of wavelike process. This process of heat transfer has been studied at great length by physicists throughout the ages, because it has been one of the most difficult and most confusing of all the questions raised by natural philosophy. It is the question of the nature of radiant energy and it embraces not only the nature of heat radiation, but also that of light, radio, X rays, cosmic radiation, and many allied phenomena to be considered later in this study.

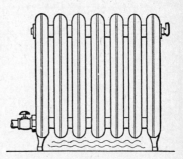

Fig. 9.5. Heating underneath a radiator is accomplished largely by radiation.

Nature of Radiant Energy. A great deal of attention has been given to the question of whether this radiation is corpuscular or wavelike. Although the present view is very complicated by virtue of certain nonmaterialistic concepts associated with quantum theory and relativity, the view has been developed that for the most part radiation is wavelike. Heat radiation, then, is the same as light radiation except that the length of the waves containing the bulk of the energy is different in each case. Generally speaking, heat waves are longer than light waves, but the speed at which they travel in empty space is the same. In fact, both kinds are liberated simultaneously, and at the same speed, by a common source because they really are both part of the same thing. Actually it is more accurate to say that a source of radiant energy radiates a whole spectrum of radiation, displaying wave properties and characterized by the lengths of the waves contained therein. A certain portion of this spectrum is recognized as heat, and other portions as light, X rays, etc. Whereas the range of visible wave lengths is recognized by the

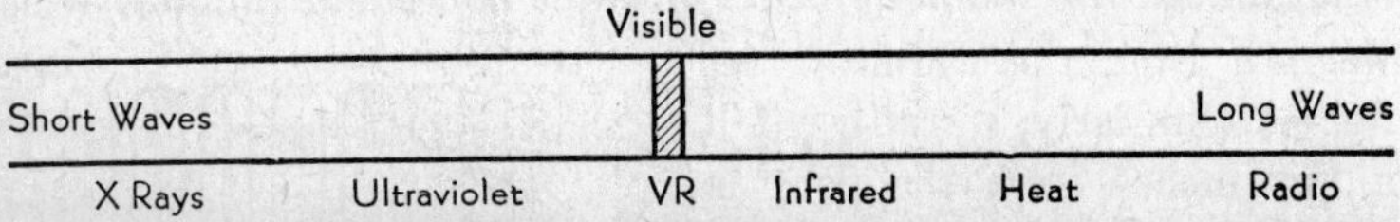

Short Waves		Visible			Long Waves
X Rays	Ultraviolet	VR	Infrared	Heat	Radio

Fig. 9.6. The electromagnetic spectrum extends all the way from the shortest X rays to the longest radio waves. The visible portion comprises a very narrow band from violet to red.

concept of color, extending from violet to red with wave lengths from 0.00004 cm. to approximately 0.00008 cm. respectively, heat waves extend from the infrared region out to the radio range where the lengths of the waves are measured in millimeters, centimeters, and even meters. Beyond the violet side of the visible spectrum the waves range from the so-called ultraviolet, responsible in large measure for sunburn and tan, to the X rays. See fig. 9.6.

Color Associated with Radiation. That color is associated somewhat with heat radiation may also be shown by calling attention to the once familiar kitchen stove. As a stove heats up, one first notices the feeling of warmth before there is any indication of color, but above a certain temperature, the stove covers glow with a dull red color. This color changes with temperature, however, becoming more yellow and perhaps finally white hot as the temperature increases, at which time it is found that a very wide range of wave lengths is being emitted. Camera film sensitized particularly for infrared is capable of detecting these radiations long before the stove becomes hot enough to glow visibly, and even in the dark. Similarly it is possible to detect the various other wave lengths present in the radiations from the white-hot covers. The analysis of all these radiations is usually referred to as *spectroscopy* and will be considered more fully in the study of light, where it seems to fit more appropriately into our logical study of physics. The study of heat is so closely tied in with the study of light, and even of radio, as to make the study of any one topic alone nearly impossible.

Law of Heat Radiation. Getting back to the study of heat radiation, we find there are certain laws governing this phenomenon. The rate of radiation depends upon the size of the radiating surface, its absolute temperature, and its nature. The dependence on temperature is not a straight proportionality but a fourth-power proportionality; i.e., a doubling of temperature, expressed on the absolute scale, does not double the rate of radiation, but increases it sixteenfold. Thus, although all bodies with temperatures above absolute zero radiate, hot bodies cool ever so much faster by radiation than cold ones.

Surface Characteristics of Radiators. The nature of the radiating surface is very important as regards the rate of radiation. Rough surfaces cool faster than smooth ones and black surfaces radiate faster than white ones. The worst conceivable radiator, then, is a perfectly reflecting surface such as a highly polished silver mirror, and it is for this reason that modern flatirons are nickel-plated all over except on the bottom, which is the only surface from which heat is allowed to escape. This reasoning suggests certain facts well known to the practical person, such as that black stoves and radiators give off more heat than those painted in light colors, and that polished copper hot-water tanks retain heat longer than painted ones. It is not just imagination, either, that makes light-colored clothes feel cooler in the hot weather than dark ones. Dark clothes actually are hotter because of radiation and its associated phenomenon of absorption. A substance has to absorb heat in order to radiate it, and so a good radiator is a good absorber. On the other hand, a woolen shirt is more comfortable than a cotton one if the wearer is exposed to the direct rays of the sun, even in the summer. This is because of the poorer conductivity of the wool, rather than its radiation ability.

The Vacuum Bottle. Thus the transfer of heat is accounted for in three distinct ways, and sometimes by all three at the same time. The problem of insulation, although mentioned previously only in connection with conduction, actually involves the elimination of all three. The success of the vacuum bottle, for example, is achieved by reducing all three modes of heat transfer to a minimum. The vacuum bottle consists of a double-walled container, whose walls are separated by a vacuum, the very worst heat conductor imaginable. See fig. 9.7. The only place where heat can escape by conduction is through the stopper and this is usually small and made of cork. The vacuum space also eliminates the possibility of convection currents. Finally, the walls are usually silvered so that whatever heat might escape across the vacuum by radiation is reflected back into the bottle. For these reasons, then, a good vacuum bottle can keep foods or liquids hot for a considerable length of time. The heat just cannot escape. Moreover, a vacuum bottle will keep cold things cold longer than it will keep hot things hot, because hot objects radiate more rapidly than cold ones.

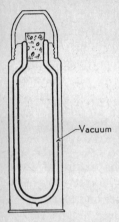

Fig. 9.7. The vacuum bottle is a double-walled glass bottle. The region between the walls is highly exhausted to prevent the transfer of heat by conduction and convection. The walls are silvered to prevent radiation.

Selective Transmission and Absorption of Radiation. That radiant heat has much the same, if not exactly the same, nature as light is also shown by the fact that it can be reflected by mirrors and focused by lenses. Of course, certain substances are more transparent to light waves (short) than to heat waves (long); i.e., they transmit waves of one length but are opaque to waves of different length. Some substances also absorb selected wave lengths. The hothouse provides an interesting situation. Light is readily transmitted through the glass windows and much radiant energy is absorbed by the objects and ground within the hothouse. Upon absorption these bodies immediately reradiate the energy as heat waves, but with lengths too great to penetrate window glass. See fig. 9.8. Thus energy enters via short waves but cannot escape because the short waves are changed to long ones. This emphasis on waves and their lengths naturally raises the question, however, as to the certainty of the view that radiant heat and light are wavelike. Furthermore, it must be admitted that no discussion of radiation would be complete without some reference to the now popular quantum theory, which contradicts the wave theory.

The Quantum Theory. About 1900, Max Planck, the German physicist, was making an intensive study of the distribution of energy in the radiation spectrum. In an attempt to formulate a mathematical relationship governing the amount of energy allocated to the different wave lengths, i.e., associated with different colors, he became convinced that the laws of classical physics were inadequate to explain observable facts. In other words, his search for the truth, which is the ultimate task of the scientist, led to the conclusion that supposedly cor-

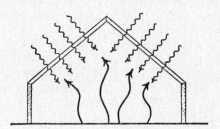

Fig. 9.8. In the hothouse short waves enter and are absorbed. They are transformed into long waves which cannot be transmitted by the glass.

rect laws were not altogether true, and that all natural phenomena are not necessarily explainable in terms of simple postulates. In particular, he found that instead of assuming radiant energy to be propagated like a continuous wave disturbance, it is necessary to postulate that energy is radiated in packets or bundles, each made up of an integral number of elementary units which he called *quanta*. This radical and very artificial postulate, offered by one of the world's most conservative physicists, was strongly resisted at the time, but today after more than two thirds of a century of the most exhaustive research in the history of physics, this quantum postulate has become the accepted view.

Some Philosophical Aspects of Quantum Theory. Even though this seems to give a corpuscular aspect to the nature of radiant energy, it has not eliminated the wave hypothesis because of other phenomena (to be discussed later in optics, pp. 206–210) which cannot be explained by corpuscles, unless the corpuscles themselves be thought of as bundles of waves. Thus radiation is still discussed in terms of waves, but also in terms of quanta, at the same time. If now this appears slightly indefinite to the student, then he has just begun to grasp the point of view of the physicist of the last generation, who felt that the discovery of the quantum theory completely undermined all semblance of reality in physics. During the present century however, many of the philosophical aspects of this whole question have been straightened out in such a way as to show that the earlier views, although approximations to the truth, are not such good approximations as the later views. It is as if the early views explained only main features but not details, whereas the present views are more comprehensive even if more complex.

The "Newer" Physics Goes beyond "Classical Physics." Here it is that we begin to see how a change has come over physics since the turn of the century, and why it is that physicists refer to the New Physics as distinguished from Classical Physics. The New Physics, more appropriately referred to as Contemporary Physics, is permeated by a point of view altogether lacking in its respect for "common sense," but characterized by strictly mathematical logic and deductions from formal assumptions justified only by their observable results. Although this is not the physics of Newton, it is nevertheless fair to point out that the major discrepancies between the two lie chiefly in the domain of submicroscopic affairs, and that most of the deductions from Newtonian physics still hold as true as ever in the ordinary mechanical world of men and machines. Today, however, we do deal with molecules, atoms, electrons, and other subatomic concepts which were not in Newton's vocabulary and which we appreciate not by our common senses, but through the eyes of the mind. Consequently quantum physics has become established in today's science, because it "explains" molecular and atomic phenomena.

Additional Discrepancies of "Classical Physics." If the quantum theory were the only source of worry to the classical physicist we should probably not be so concerned about it; but the facts are that many of the supposedly minor discrepancies of classical physics have become grossly magnified by the researches of modern times; and, strangely enough, those measures required to restore order and reason to the troubled picture involve a break with common sense in much the same way that is exemplified by the quantum theory. Specifically, we refer to the study of relative motion, which Einstein has explained only by reference to concepts and mathematical logic which do not seem sensible from the earlier point of view. One of these, for example, is the relativity postulate that no body can have a velocity greater than that of light. In brief, the modern physicists, attempting to understand nature, have discovered that instead of being faced simply with the task of polishing off certain rough spots of otherwise completely solved problems, they have actually scratched the surface of much larger and profoundly more significant problems. There is, however, no reason to be discouraged, because the discoveries of the past have been very well consolidated, and it is confidently expected that what will be discovered in the future will somehow be reconciled with present knowledge. Much of the new physics deals with electrical phenomena, the elementary concepts of which we shall consider in the next chapter.

Summary. In this chapter the three modes of heat transfer—conduction, convection, and radiation—have been described. Insulation is seen to be the inverse of conduction. Conduction refers to the transfer of molecular energy right through an object. Convection involves the circulation of heat by some agent which actually carries it from one place to another. Radiation is a wave-like propagation quite different from the other two modes. It is electromagnetic in nature like light, radio, and X rays. It is best explained by the quantum theory, which has almost completely revolutionized the point of view of physics, so that modern physics is something more than an extension of classical physics, although the latter is a first approximation of the former.

QUESTIONS

1. Some people were inclined to boast about the length of time they could keep ice in a refrigerator by completely wrapping it in newspaper. Comment.
2. Why is a thick layer of air used for insulating ovens?

3. Icicles are sometimes observed to melt on the south side of a house before they do on the north side. Why?
4. What is the effect of a fireplace on the ventilation of a room?
5. Why may a fireplace which smokes at first, stop smoking after a fire has burned a short while?
6. Explain how it is that a person in front of a fireplace fire may have a cold back yet feel hot in the face.
7. Why does a thermos bottle keep cold things cold longer than it can keep hot things hot?
8. What are some of the features of the quantum theory?

CHAPTER X

ELECTRICAL CONSIDERATIONS

STATIC ELECTRICITY, CHARGES, POTENTIAL, CAPACITY

In an earlier chapter reference was made to the electrical structure of matter. It was pointed out that all matter is constructed fundamentally from charges of electricity. The concept of electrical charge, although a somewhat abstract concept, is so important that it cannot go unmentioned in this description of the physical world. In the beginning, the concept was just a device invented to describe a situation, but later it was given reality by the discovery of natural units of charge. Today it is felt that such units of charge as the electron, the positron, the proton, the various mesons, etc. really exist and are fundamental entities basic to the very concept of matter itself. Hence they invite our study.

Frictional Electricity. Proceeding historically and logically to the development of electrical knowledge, we start with a consideration of frictional electricity and electrostatics. As was pointed out earlier, if a rod of hard rubber or ebonite is rubbed vigorously with fur it becomes endowed with the peculiar ability to pick up bits of paper and to attract small pieces of pith to itself. See fig. 10.1.

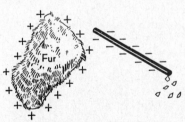

Fig. 10.1. An ebonite rod which has been rubbed with fur will pick up bits of paper due to its electrical charge.

Similarity between Electric and Magnetic Forces. The force thus demonstrated has been called an electrical force after the Greek word for amber, the substance for which it was first observed. A similar but quite different force has also been observed in connection with iron filings and the ore magnetite, or lodestone, which was used at an early date as a mariner's compass because of its directional properties. See fig. 10.2. This force has been called a magnetic force. See chapter XI for a more complete discussion of magnetism. In each of these two cases the force is unique because, unlike ordinary pushes and pulls requiring contact between bodies for their effects to be felt, no contact is required. Early physicists described such phenomena as action at a distance. The force of gravity being the only other such force met with in physics, the three together constitute a special group. For reasons which will become apparent later, these are called "field" forces in contrast to "contact" forces. Although they all display certain common aspects they are nevertheless quite different and relate to entirely distinct phases of physics. It is found however, as will be shown presently, that electrical and magnetic phenomena are related to each other so that it is natural to consider these together, apart from the study of gravity.

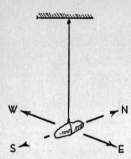

Fig. 10.2. An elongated piece of magnetite, or lodestone, if suspended by a string will point toward the north.

Two-Fluid Theory of Electricity. Returning to electrical phenomena, it is supposed that electric forces are produced by so-called *electric charges* which can be accumulated upon, or wiped off from, bodies when they are rubbed with other bodies, as in the case of the ebonite rod rubbed with the fur. It is found that an ebonite rod that has been thus rubbed with fur tends to repel another ebonite rod similarly treated, but that it tends to attract toward itself a glass rod that has been rubbed with silk. It is as if the charges on the ebonite and the glass were quite different—in fact, opposite in character. Also, the fur is found to attract the ebonite, and the silk to attract the glass after they have been thus rubbed together. These effects can better be observed by charging small pith balls hung from silk threads. Charging is accomplished by bringing the pith balls in contact with the charged rods of ebonite and glass respectively. Such observations as these gave rise to the early view that there were two distinct kinds of electric charges, indicated for convenience by the names "positive" and "negative." All those substances which, when rubbed with other substances, behaved like the glass rod when rubbed with silk were said to be charged *positively;* and all those behaving like the ebonite rod when rubbed with fur were said to be charged *negatively*. The matter of which was positive was purely arbitrary, and so too much emphasis must not be placed upon the physical significance of the terms "positive" and "negative," except that the one is the opposite of the other. Thus two fundamental kinds of electricity were postulated and a very important law was discovered which states in part that like charges repel and unlike charges attract each other. See p. 130 for details.

One-Fluid Theory. Benjamin Franklin, the American statesman and scientist, feeling that this so-called two-fluid theory of electricity was unnecessarily complicated, proposed a view which in many ways resembles the modern viewpoint. He proposed a so-called single-fluid theory, postulating that the positive charges represented an accumulation of charge (presumably positive), whereas the negative charge was merely a deficiency of such a positive charge. Except that the modern electron theory involves charges inherently negative instead of positive, i.e., the kind exhibited by the ebonite rather than the glass in the preceding paragraph, it explains ordinary electrostatic phenomena in a similar way, i.e., by a consideration of surpluses and deficiencies of a single kind of electric charge.

Electrical Terminology Based upon Positive Fluid Theory. One of the difficulties encountered in the study of electricity today is that, although the negative electron theory has superseded the positive fluid theory just mentioned, the older terminology still persists, and many electrical concepts are actually defined in terms of the positively charged fluid of Franklin. Thus even the unit of electrical charge, arbitrarily defined for purposes of a logical understanding of electrical phenomena, is not the electron, because this natural unit was not discovered until about the beginning of the present century, but the so-called *unit positive charge*. (See definition, p. 130.)

Abstract Nature of the Study of Electricity. This important concept of the unit of charge is in reality only a product of the mind. It is purely supposititious, but its appreciation is absolutely necessary for the proper understanding of all electrical concepts, since the science of electricity has been developed as an abstract subject. Electricity, unlike mechanics, for example, deals with concepts alone rather than visible things, and the student who wishes to understand a given electrical situation must, without exception, follow through a train of carefully developed logical arguments couched in precisely defined terms, because there are no revolving wheels or bouncing springs to watch here. It is relatively easy to discuss academically the motion of a body under certain circumstances, since bodies in motion are directly observable, and the arguments can be checked by direct observation. One does not make direct observations upon electric charge, however. All observations here are of the indirect type which requires an understanding of the relationships involved, because we do not have direct sense perception in electricity. Electric charge cannot be seen, felt, heard, smelled, or tasted. Hence the physicist is particularly careful about his definitions of electrical terms.

Coulomb's Law. The unit positive charge, or the unit of electricity as it is popularly called, is defined in terms of a certain law called *Coulomb's Law*. Not only do like charges repel and unlike charges attract each other, but the magnitudes of these forces is found to depend upon the amounts of charge and the separation between them. Coulomb's law states this in mathematical language by an expression closely resembling Newton's law of gravitation (which, it will be recalled, is a so-called inverse square law), because the force between charges diminishes as the second power of the separation. The law states that the force between two charges varies directly as the product of the charges and inversely as the square of the separation.

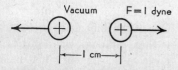

Fig. 10.3. If two pointlike charges placed one centimeter apart in vacuum repel each other with a force of one dyne, they each represent the electrostatic unit of charge called the statcoulomb.

In accordance with this law, *the unit of charge* is defined as that amount of electricity which, placed one centimeter away from a like amount in vacuum, repels with the force of one dyne. See fig. 10.3.

Since the dyne is that force which will give a gram mass an acceleration of only one centimeter per second per second, the

8. A positively charged rod is brought near a charged electroscope. If the leaves collapse, what is the charge on the electroscope?
9. Is there any point in lying flat on the ground if one is caught out in the open in a severe thunderstorm? Should one stand under an isolated tree?
10. How does the mass of the nucleus of a hydrogen atom compare with that of an electron?
11. If two charged ping-pong balls repel each other, what can be said about their charges?

charged without coming into contact with the charged body; moreover, the sign of the induced charge is the opposite of that on the original charged body. See fig. 10.9.

Summary. The concepts of charge, potential, and capacity or capacitance are so closely interrelated that no satisfactory description of electrical phenomena can be given without referring to them. That is why so much attention has been devoted to them in this text. Although such attention may not seem necessary for any three particular concepts in a descriptive survey, the study of electricity is so abstract that considerable emphasis has to be laid upon the basic considerations. Otherwise the later study of electric currents and allied phenomena would be a mere display of words. With a proper basic vocabulary, on the other hand, such material as will be considered in later chapters can be very significant even if its consideration is restricted to the descriptive level. Before studying electric currents, however, it seems fitting that we should turn our attention next to a subject very closely related to electricity and already referred to earlier, namely magnetism. In the next chapter magnetic phenomena will be considered and the subject of magnetism will be developed in much the same manner as its sister subject, static electricity.

QUESTIONS

1. What is meant by the term "capacitance"?
2. What causes a small pith ball which has been attracted to a charged body to be suddenly repelled from the charged body after coming in contact with it?
3. The dangling chain formerly observed on the gasoline truck grounded the truck. Explain how this minimized the risk of fire.
4. Why is it recommended that the cleaning of silk garments in gasoline be done out of doors?
5. How may the "leaking" of charge from pointed metallic objects be explained?
6. Why is it better not to say that a positively charged object has gained positive charges, but rather to say that it has lost electrons?
7. An uncharged pith ball is suspended by a string. What happens to it when a positively charged rod is brought near it?

rubbed with silk, then the leaves will remain separated with this charge upon them. Now a positively charged body brought up to the electroscope will cause the leaves to diverge farther, but a negatively charged body will cause the leaves to collapse more or less. Thus an unknown charge can be compared with a known one.

Charging by Induction. It is possible to transfer a charge to an electroscope and leave it there without the electroscope being touched by the charged body, i.e., by *induction*. This is accomplished by first bringing the charged body near the knob of the

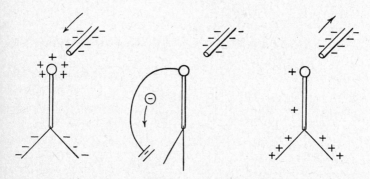

Fig. 10.9. Charging an electroscope by induction. The charge left on the electroscope is of opposite sign to that on the charging rod.

electroscope, being careful not to actually touch it, whereupon the leaves will diverge. This is due to the fact that the charge attracts to the near end of the knob of the instrument charges of opposite sign and repels to the gold-leaf end of the instrument an equal number of charges of the same sign, but without adding to or subtracting from the total amount of charge on the neutral electroscope because no contact has been made. By definition a neutral body has an equal number of positive and negative charges. If, while the charged body is thus held near the electroscope, the latter is grounded, the leaves will collapse because the ground furnishes whatever charge is needed to neutralize the influence of the charging body, and relieves the electroscope of the responsibility. If now, after breaking the ground connections, the charged body is removed, the leaves will again diverge because the extra charge supplied by the ground while the charged body was nearby has no place to go and is not needed either. It constitutes an extra charge on the electroscope, which has thus been suggest the origin of the term condenser, whose function is, so to speak, to condense electricity, i.e., to concentrate it, because two conductors separated by an insulator can accommodate more charge at the same difference in potential than either conductor alone with respect to ground.

Detection of Electric Charge—The Gold-Leaf Electroscope. This discussion would not be complete without some mention of the manner in which electrical charges are detected. The so-called gold-leaf electroscope is the instrument most commonly used for this purpose. It consists simply of a narrow strip of gold foil hung over a hook suspended through the insulated stopper of a glass container in such a manner that the two ends of the foil hang down inside the glass without touching it. The glass container serves merely as protection against outside influences such as air cur-

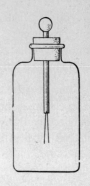

Fig. 10.7. Diagram of a gold-leaf electroscope enclosed in a container to shield it from air currents.

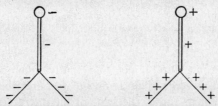

Fig. 10.8. When charged either positively or negatively the gold leaves diverge.

rents. The top end of the hook terminates in a metal ball or knob on the outer side of the stopper. See fig. 10.7. When a charged body is touched to the knob of the electroscope, the gold leaves diverge because they each receive the same kind of charge and are mutually repelled. Of course the sign of the charge cannot be directly detected because the leaves will diverge upon receiving either kind of charge. See fig. 10.8.

Indirectly, however, it may be possible to determine the sign of a charge, as follows: If first the knob is touched with a charge of known sign, such as the positive charge on glass which has been case. The latter might be said to have more staying power. Put differently, it is obvious that in the case of the second pipe more water is required to raise the pressure a given amount than in the case of the former, because more water is required to raise the surface level a given amount.

Electrically, a similar situation holds. Conductors of large dimensions require more charge to raise the potential by a given amount than do small conductors. See fig. 10.6. Conductors thus have a characteristic known as *capacitance* or sometimes just called *capacity,* by which term is meant, not the amount of electricity that the conductor can hold, but rather the amount necessary to raise the potential by a unit amount. One stat-coulomb per stat-volt is called a *stat-farad* of capacity whereas one coulomb per volt is called a *farad,* in honor of Michael Faraday. A conductor or a combination of conductors and insulators having capacitance, or capacity, is called a *condenser,* or a *capacitor.* Condensers consisting of layers of tin foil separated by layers of waxed paper and packed into sealed containers are widely used in all telephones, radios, automobiles, and innumerable other electrical devices, and so play a very important role in modern life.

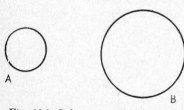

Fig. 10.6. Sphere A will develop a greater potential than B with the same amount of charge, owing to the larger capacitance of B.

An Application of the Condenser. One use of condensers (capacitors) is to be found in the automobile. The function of the spark plug is to produce sparks between two electrical terminals at proper intervals. As has been suggested above, the spark will occur when the potential difference between the electrodes is raised to the proper value. If there is a condenser in the spark plug circuit, the amount of charge required to raise the potential to the proper value is increased, with the result that when the spark does occur a greater charge and therefore a greater amount of energy than otherwise will be released and the spark will be more intense, or hotter, as the garage man would say. This may the silk, electrons are wiped off the glass and crowded on the silk. Reference is frequently made, however, in quantitative discussions to the old unit positive charge, but this is simply to be thought of as the absence of about two billion electrons. Of course, the charging of a body by friction in no way involves the generation of charge since matter itself is made of charges. Charges are not created but merely separated by such processes. Potential differences, on the other hand, are created as charges are separated.

Electric Capacity or Capacitance. There is still another aspect to this situation. A given amount of charge does not always establish the same potential for every body any more than a given amount of water produces the same pressure at the base of its container. Since hydrostatic pressure at a given place depends upon the depth of the water there, a distinction must be made between the pressure developed by the same amount of water in two different containers of different heights and correspondingly different cross sections. Obviously the pressure at the base of a tall slender standpipe is greater than that at the base of a short pipe of correspondingly larger diameter holding the same volume of water. See fig. 10.5.

On the other hand, if water is released from these two pipes simultaneously, the pressure will drop more rapidly in the former

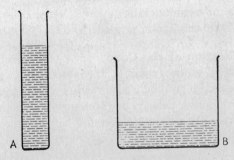

Fig. 10.5. Although the two tanks contain the same amount of water, the pressure at the base of A is greater than the pressure at the base of B. Electrically, tank B would be said to have more capacitance; i.e., it requires more material to increase the pressure by unit amount than does tank A.

as an important one, considering that opposite charges just naturally try to get together and resist separation. Here the matter of insulation is the key to the situation. All substances can be listed in order of their ability to conduct electrical charges. Those at the top end of the list are referred to as *conductors* and those at the bottom end as *insulators*. The metals, in general, constitute the former group, and substances like porcelain, glass, hard rubber, ebonite, etc. constitute the latter. Dry air is also a good insulator; as much as 30,000 volts difference in potential can be maintained between spherical conductors one centimeter in radius and one centimeter apart in dry air.

Production of Potential Differences. The production of a difference in potential is a matter of separating charges. This is just what happens when the ebonite rod is rubbed with fur. Electrical charges are simply separated by friction, the rod becoming charged negatively and the fur positively. But let us now ask where the charges originate to be thus capable of separation by friction; i.e., let us consider the electron theory of the structure of matter.

The Electron Theory. As was previously mentioned, the unit positive charge is not a natural unit even though it may once have been so conceived by early physicists. It was an invention designed to facilitate the study of electrical phenomena. During the last decade of the 19th century J. J. Thomson had discovered evidence for the existence of a naturally occurring unit of negative electricity (the electron). See also p. 165. This view became established when the American physicist Millikan about 1910 succeeded in measuring the charge on the electron in terms of the unit positive charge. It was found to be negative and very small, less than one-half a billionth of a stat-coulomb ($e = -4.80 \times 10^{-10}$ stat-coulombs).

Without going into the details of this discovery at this time, we shall treat the remainder of this description of electrical physics in terms of the now accepted electron theory, which states that most electrostatic phenomena are due to the existence of free electrons associated with matter. Thus it is believed today that when a rod of ebonite is rubbed with fur, what really happens is that negative electrons are removed from the fur and accumulated on the rod. Similarly in the case of the glass rod and is capable of moving it back toward the negative charge, if released, just as an unsupported book or other object is capable of falling to the floor, under the action of gravity. In the vicinity of a positively charged body, work must be done upon a positive charge, or a positively charged body, to move it from one point to another point nearer the positive charge. This is described by saying that points in the vicinity of positive charges are at relatively *high potential*. By the same argument it follows that points near negative charges are at *low potential*. Thus the *difference in potential* between two points represents the amount of work necessary to move a unit amount of charge from the one to the other, just as the difference in gravitational potential between two levels represents the amount of work necessary to raise a unit mass from the lower to the higher level. If the unit amount of charge is one stat-coulomb then the difference in potential is one *stat-volt* if the work done is one erg. The volt, which is 1/300th of a stat-volt, has been adopted as the *practical* unit of potential. It is also the M.K.S. unit of potential. In popular discussions voltage and potential are terms often used synonymously, if not always correctly. The volt is the difference in potential if one joule of work is required to move one coulomb of charge.

Mechanical Analogy of Potential Difference. Regarding the question of danger associated with high potential, or more correctly, with large differences in potential, the situation is analogous to that of a missile raised to a high elevation. If after being raised it should be released, energy would be liberated in an amount depending upon the mass of the missile as well as the elevation of it. So with electricity; a large difference in potential represents a store of electrical potential energy. If it should be released by the introduction of a conducting path between the two points experiencing this difference in potential, then energy would be liberated in an amount depending upon the magnitude of the charge and the difference in potential. This is precisely what happens in a flash of lightning, which is nothing but a release of electrical energy. Potential difference is often referred to rather loosely as electrical pressure, because of the analogy with hydrostatic pressure.

Electrical Conduction and Insulation. The question of how a difference in potential is maintained is an interesting as well

"Intensity of the Electric Field," which is identically the same thing. This vector quantity is usually labelled "E."

Considerations of the field intensity (E) at different points in the field of a charged body suggest the concept of *lines of force* to represent this quantity (Fig. 10.4). A line of force is the path that

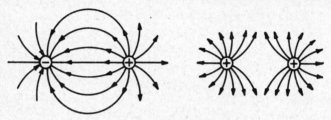

Fig. 10.4. Electric lines of force.

would be taken by a free unit positive charge, and it is customary to imagine lines of force emanating from positive charges and entering negative charges (a unit positive charge being repelled from a positive charge and attracted toward a negative charge). The magnitude of E at a given point is represented by the concentration of lines of force at that place, measured in number of lines passing perpendicularly through unit area of surface.

Electric Potential. A common term in electrical language is *potential*. One often hears references to high-potential wires and the dangers associated with them. Sometimes it is called high voltage. What does this mean and why is voltage so important in this electrical age? What is voltage anyway?

The answers to these questions are quite definite now that the concept of the electric charge has been established. Since a positive charge, for example, exerts an attractive force on a negative charge, a force must be exerted to separate them. If they are to be separated work must be done in the process, because work is always done when a body is displaced in the same direction as that of a force applied to it. Similarly, work must be done to push two charges of like sign together. This is equivalent to saying that electrical potential energy is involved in the process. Specifically, when a positive charge is moved away from a stationary negative charge, or a negatively charged body, work must be done upon it and it is said to acquire electrical potential energy. This energy electrostatic unit of charge just defined is a rather small quantity. This is the C.G.S. unit of charge and is called the *stat-coulomb*. Incidentally it will be noted that all electrical units are named after famous physicists, but with certain prefixes attached to indicate the system to which they belong, since there are several systems of electrical units. In M.K.S. units, the *coulomb* is the unit of charge. It is equivalent to 3×10^9 *stat-coulombs*.

Concept of Electric Field. Thus the concept of electric charge is established as a sort of speck of electricity capable of exerting an influence upon all other such specks in its neighborhood. The regions around charges are therefore places where electrical forces are manifested; such regions are usually called *electric fields*.

Electric Field Intensity. What is now more natural than to gauge the electrical intensity of a field, if only in the mind's eye, by the reaction of a unit positive charge to it? This gives rise to the important concept of *electrical field intensity* at a point. It is defined as the force per unit positive charge at the point in question and it is a vector quantity. It is perhaps becoming clear to the student that the study of electricity, with its abstract concepts, is developed as a study of the activity and behavior of electric charges, with the unit positive charge used for exploring and testing purposes, and that the procedures involved are imagined to take place rather than directly observed.

The field concept, whether it be the electric field or the magnetic field to be described in the next chapter, or for that matter the gravitational field, carries additional significance to the theoretical physicist. It is one thing to think of the region surrounding a charge as the field due to that charge so that a unit positive charge introduced into it will experience a force of attraction or repulsion as the case may be. The theoretician, however, finds it advantageous to attribute the force on the unit charge in a field to the existence of the field itself, i.e., to assume that a field is characterized by a property by virtue of which a force acts on a charge located in it. In other words, a field exists between two oppositely charged plates without reference to whether or not it is the region surrounding a single charge or a single charged body. In this context reference is usually made to "The Electric Field" (a vector quantity) rather than to the

CHAPTER XI

MAGNETISM

Elementary Magnetic Phenomena. Everyone probably has had some contact with magnetism, if only a superficial one. Possibly this involved a toy horseshoe magnet and led to an appreciation of the simple fact that a common horseshoe magnet will attract steel pins and iron filings but will not attract any-

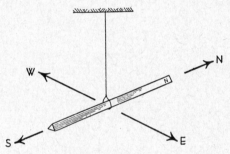

Fig. 11.1. A slender bar magnet, like a piece of lodestone, will point north if suspended by a string as shown.

thing made of brass. It is also perhaps well known that a slender bar magnet suspended transversely by a thread tied around it at the center will so orient itself as to point in a north-south direction. See fig. 11.1. This observation was made by the ancients, who suspended elongated samples of the ore magnetite, commonly called lodestone, and thus the mariner's compass was invented, supposedly by the Chinese thousands of years ago. Here, as in our study of electrostatics, the question is again asked concerning the nature of these phenomena.

It is also noticed that the opposite ends of a bar magnet act as though they are magnetized oppositely. Because of the directional aspect, the terms north and south are used instead of positive and negative to designate the two ends. It is also noticed that the north ends of two different magnets repel each other but that a north end of one attracts a south end of another, just as like electric charges of electricity repel and unlike charges attract each other.

Explanations of Magnetism. These phenomena suggested to early physicists that magnetism, like electricity, was corpuscular in nature, involving small units of magnetism analogous to units of electric charge. Such units were called *magnetic poles*, north and south, rather than positive and negative, in view of the association of magnetic phenomena with geographical north and south poles. A law of attraction and repulsion of these magnetic poles was also discovered by Coulomb stating that like poles repel and that unlike poles attract with a force which is directly proportional to the products of the pole strengths and inversely proportional to the square of their separation just as in the analogous electrical case. The *unit magnetic pole* was also defined in a manner similar to that employed to define the unit charge. Today, however, no special significance is attached to this unit pole, not only because no such natural unit of magnetism has ever been discovered, but because it seems to be possible to explain magnetic phenomena in terms of moving electric charges, making the pole concept unnecessary. Nevertheless, the pole concept persists in magnetic terminology and is still frequently used in elementary discussions of magnetic phenomena. However, following the discovery about 1819 by the Danish scientist Oersted that a pivoted compass needle was deflected when it was placed near an electric charge-carrying wire, the pole concept has become increasingly obsolete. See p. 157 for a discussion of the magnetic effect of current.

Terrestrial Magnetism. The earth's magnetism provides a very interesting topic for study. It appears that the earth behaves like a large spherical magnet with a north-south axis which makes a relatively small angle with its geometric axis. The north end of a compass needle does not point true north, i.e., in the direction of the north star, but toward the magnetic pole, which is situated in the general vicinity of Hudson Bay in Canada. This pole of the compass is sometimes referred to as the north-seeking pole. Also, since the geographic and magnetic poles of the earth do not coincide, it follows that compass needles which presumably point north are likely to be in error at any given point on the earth's surface. In New England, this error, or *angle of declination* as it is called, amounts to nearly twenty degrees west.

Angle of Dip. Moreover, the supposititious magnetic poles of the earth are not to be thought of as located upon the surface of the earth, but as at a considerable depth. This is to account for the fact that a compass needle suspended so as to be free to move up and down as well as from right to left tends to display a downward tendency at all points except on the magnetic equator. See fig. 11.2. At the poles the needle points straight up and down. The angle thus made with the horizontal is called the *angle of dip* and in New England amounts to some seventy-odd degrees. Thus in New England the downward tendency outweighs the horizontal one by three to one. Technically it is said that the vertical component of the earth's magnetic field is three times as great as the horizontal component in this vicinity. As in the study of electrostatics, the *intensity* of a *magnetic field* at a point is attributed to the existence of a neighboring magnetic pole. It serves to measure the pole strength of the magnet, which to the layman is a measure of the strength of the magnet in question.

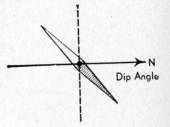

Fig. 11.2. A compass needle mounted so as to be free to rotate in a vertical plane will dip below the horizon in the northern hemisphere.

Magnetic Lines of Force. If the region around a bar magnet is explored with a small compass it is found that the north end of the compass always points away from the north pole and towards the south pole of the magnet at every point. Imaginary lines are suggested that indicate the direction that a supposititious north magnetic pole would follow if free to move in the region. These are called *lines of force*. (Fig. 11.3.) They emanate from the north end and enter the south end of the bar magnet, and provide a way of visualizing the direction of magnetic forces where such forces exist.

Magnetic Materials. It is customary to associate magnetic phenomena with the substance iron. This is due not only to the early discoveries of magnetic properties among ores of iron, but

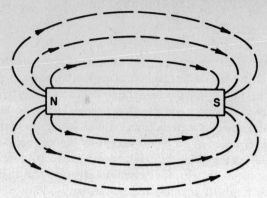

Fig. 11.3. Magnetic lines of force.

also to the fact that iron is the most conspicuous of the naturally occurring magnetic substances. It would be a mistake, however, to assume that iron is the only magnetic substance. Nickel and cobalt display magnetic properties, and recently alloys have been made which have far exceeded iron in magnetic properties. One of these alloys is known as "permalloy" and was developed in the research laboratories of the American Telephone Company for use in telephone equipment. Another one of these alloys is known as "alnico."

Magnetic Induction. One of the most interesting of the various magnetic phenomena is *magnetic induction*. When a bar of unmagnetized iron is brought near one end of a magnetized bar, magnetic poles are induced in it. That is to say, the bar becomes magnetized by induction with the near end acquiring a pole of sign opposite to that of the inducing pole, and the far end acquiring a pole of the same sign as that of the inducing pole. It is as if the unmagnetized bar were composed of myriads of elongated molecular magnets capable of orienting themselves at will. In the unmagnetized state these molecules are to be thought of as oriented at random, but in the vicinity of a strong magnetic north pole, let us say, all the south ends of the molecular magnets are drawn toward the north pole and the north ends are repelled from it, causing the bar to become magnetized with equal and opposite poles. This is the essence of one theory of magnetism, which, although already out of date, still serves to provide an elementary picture. It does not completely clarify the nature of the molecular magnets, nor explain why only iron and a few other substances display this peculiar property to any marked degree. Yet for elementary purposes it is reasonably satisfactory.

Magnetic Permeability. It is by induction that most magnets are made. Bars of unmagnetized iron or other magnetic substances are subjected to strong magnetic fields which establish the magnetism by influence, so to speak. One of the features of permalloy, previously mentioned, is its ability to become very strongly magnetized under the influence of a relatively weak magnetic field. This is due to a property technically called *permeability* which affords one of the most useful means of classifying magnetic substances. If, for example, a substance is strongly magnetic its permeability is very high and the substance is said to be *ferromagnetic*. If, by its presence, a substance enhances a magnetic field but not so strongly as a ferromagnetic substance, it is called *paramagnetic*. The substance bismuth, on the other hand, is conspicuous because it weakens a magnetic field by its presence. It is said to be *diamagnetic*. Paramagnetic substances are characterized by permeabilities greater than unity, whereas a diamagnetic substance has a permeability less than unity.

Summary. Much of the study of magnetism deals with electromagnetism, which logically cannot be considered until electric currents have been studied. Such material will be the subject for consideration in the next chapter. Here we have dealt with so-called static magnetic phenomena as the previous chapter dealt with static electrical phenomena. Such a preliminary study develops the vocabulary and fundamental concepts upon which the later study is so dependent. We shall see presently how the two topics electricity and magnetism blend together in the study of electric currents.

QUESTIONS

1. What is meant by the expression "north-seeking pole"?
2. Distinguish between angle of declination and angle of dip.
3. What is meant by a magnetic field?

REVIEW QUESTIONS
(See page 217 for answers.)

Chapters VIII, IX, X, and XI

1. The scientist who suggested that heat is energy rather than a fluid was: 1. Newton; 2. Watt; 3. Gibbs; 4. Maxwell; 5. Rumford .... ( )
2. That temperature on the Centigrade scale which corresponds to 68° Fahrenheit is: 1. 20° C.; 2. 37.7° C.; 3. 68° C.; 4. 64.8° C.; 5. 55.5° C. .... ( )
3. Pyrex glass is useful for baking dishes on account of its: 1. small expansion coefficient; 2. small specific heat; 3. large specific heat; 4. large thermal conductivity; 5. large expansion coefficient .... ( )
4. The amount of heat required to raise the temperature of one pound of water by one degree Fahrenheit is called: 1. the heat of fusion; 2. the B.T.U.; 3. the calorie; 4. the erg; 5. the heat of vaporization ..... ( )
5. Franklin's pulse glass demonstrates: 1. specific heat; 2. the mechanical equivalence of heat and energy; 3. boiling under reduced pressure; 4. regelation; 5. surface tension .... ( )
6. When ice at 0° C. changes to water at 0° C.: 1. there is no heat change; 2. heat is given off by the ice; 3. heat is absorbed by the ice; 4. the volume becomes greater; 5. there is no change in volume .... ( )
7. As the atmospheric pressure is raised, the boiling temperature of a liquid is: 1. lowered; 2. raised; 3. unaltered; 4. just 100° C.; 5. 212° F. .... ( )
8. The phenomenon of melting under pressure and refreezing when the pressure is relieved is called: 1. sublimation; 2. evaporation; 3. refrigeration; 4. regelation; 5. saponification .... ( )
9. The sling hygrometer is used for measuring: 1. relative humidity; 2. atmospheric pressure; 3. the boiling point of water; 4. the freezing point of water; 5. none of the above .... ( )
10. The reason for insulating a refrigerator is: 1. to prevent outward radiation; 2. to prevent heat from coming in; 3. to prevent the cold from coming out; 4. to melt the ice; 5. to prevent the ice from melting .... ( )

REVIEW QUESTIONS

11. It is possible to heat a greenhouse by the sun because: 1. glass transmits the short wave length more readily than the long; 2. glass transmits all wave lengths equally well; 3. glass transmits long wave lengths more readily than short wave lengths; 4. glass is a good conductor of heat; 5. glass is a poor conductor of heat .... ( )
12. If the temperature of a piece of metal changes from 300° Abs. to 600° Abs., the rate of heat radiation is: 1. halved; 2. doubled; 3. tripled; 4. quadrupled; 5. increased by a factor of 16 .... ( )
13. Cork with respect to copper is a much: 1. better radiator of heat; 2. poorer insulator of heat; 3. poorer conductor of heat; 4. poorer radiator of heat; 5. better conductor of heat .... ( )
14. Heat is transferred from a steam radiator to a point in a room directly beneath it largely by: 1. conduction through the air; 2. convection currents established in the air; 3. radiation through space; 4. surface tension; 5. molecular bombardment ...... ( )
15. The quantum theory was first proposed to explain: 1. radiation phenomenon; 2. conduction; 3. convection; 4. melting; 5. freezing .... ( )
16. When an ebonite rod is rubbed with fur: 1. electrons are generated; 2. electrons are accumulated by the ebonite rod; 3. electrons are accumulated by the fur; 4. protons are acquired by the fur; 5. electricity is generated .... ( )
17. Coulomb: 1. discovered the gold leaf electroscope; 2. discovered a law of electrostatic attraction; 3. found that the electron has a negative charge; 4. observed that ebonite rubbed with fur acquires a positive charge; 5. invented the lightning rod ...... ( )
18. If two gilded ping pong balls repel each other: 1. one must be positively charged and the other negatively charged; 2. both must be charged negatively; 3. one may be neutral; 4. both may be neutral; 5. they must be charged with the same sign .... ( )
19. When an electroscope is charged by induction: 1. its charge is always negative; 2. its charge is always positive; 3. its charge has the same sign as that of

the source; 4. its charge has the opposite sign to that of the source; 5. its electron content is unaltered .... ( )
20. The leaves of a charged gold-leaf electroscope stand apart: 1. because the sign of the charge is the same on each leaf; 2. because the signs are opposite; 3. because charges remain on the outsides of conduction bodies; 4. because of electrostatic attraction; 5. because electrons have the opposite sign to protons .... ( )
21. The purpose of the dangling chain on the gasoline truck is: 1. to prevent positively charged protons from accumulating on the truck; 2. to allow protons to leave the truck at will; 3. to insure that sparking, which might start a fire, will take place at some distance from the gasoline; 4. to warn the driver if the truck should become charged and thereby increase the fire hazard; 5. to prevent an excess or deficiency of electrons from accumulating .... ( )
22. Lightning: 1. never strikes the same place twice; 2. never strikes steel-framed buildings like the Empire State building; 3. seldom strikes isolated tall objects; 4. is seldom likely to strike buildings with lightning rods only if they are well grounded; 5. is less hazardous to a building with lightning rods regardless of whether or not they are well grounded ... ( )
23. The capacitance of a condenser equals: 1. charge/current; 2. charge × potential difference; 3. charge/potential difference; 4. charge × current; 5. potential difference/charge .... ( )
24. A condenser is used to: 1. create charge; 2. annihilate charge; 3. store charge; 4. create energy; 5. destroy energy .... ( )
25. Electricity is: 1. a form of matter; 2. a manifestation of radiation; 3. a basic concept; 4. a form of energy; 5. explainable in terms of mass, length, and time .... ( )
26. A compass needle points in a north-south direction: 1. because it is electrically charged; 2. because the earth is electrically charged; 3. because the earth is a magnet; 4. because the magnetism of the earth is uniformly distributed; 5. because north poles repel south poles .... ( )

27. The angle between the true geographical north and the magnetic north is known as the angle of: 1. inclination; 2. deviation; 3. error; 4. declination; 5. dip .... ( )
28. The angle of dip refers to: 1. the angle between true north and magnetic north; 2. the angle made by the compass needle with the horizontal; 3. the angle of declination; 4. the ecliptic; 5. the angle of latitude .... ( )
29. A magnetic field produced by a horseshoe magnet is directed: 1. toward the north pole of the magnet; 2. toward the south pole; 3. away from the south pole; 4. has nothing to do with the two poles; 5. none of the above .... ( )

CHAPTER XII

ELECTRICAL CONSIDERATIONS (*Continued*)

CURRENT ELECTRICITY

This is an electrical age, and yet the number of people who still seem mystified by the subject of electricity is relatively large. Part of the difficulty is undoubtedly due to the exacting manner in which electrical and magnetic terms are defined, as has been brought out in the two preceding chapters. However, the student who has followed this presentation thus far should find himself prepared for a descriptive consideration of those phenomena which make this an electrical age—an age of electric currents, electric power, electrical devices, and allied topics. The present chapter will deal with electric currents.

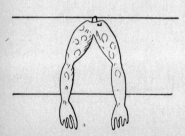

Fig. 12.1. A pair of frog's legs hung from a wire was observed by Galvani to twitch as contact was accidentally made with a second wire made of different material from the first wire.

Galvani's Experiment. The Italian scientist Galvani is credited with being the first to have discovered the existence of currents of electricity. He is said to have hung a pair of frog's legs from a metal wire near which ran a second wire of different material connected with the first. See fig. 12.1 As the frog's legs accidentally touched the second wire they were observed to twitch. The interpretation given to the phenomenon was that a difference in electrical potential, established chemically through the frog's legs and the dissimilar metals, caused electricity to flow through the circuit thus established.

Flow of Charge. It is clear from earlier considerations that if a conducting path is provided between two points charged to different potentials, electrical charge should pass from one point to the other in an attempt to equalize the potential. This would ordinarily be just a single surge of charge. But if some arrangement could be provided to maintain a difference in potential between the two points while at the same time charges were passing, then a so-called current of electricity would be established. This suggests, of course, a generation of potential difference at a rate great enough to offset the dissipation allowed by the conducting path. The situation is entirely analogous to the flow of water out of a reservoir under a certain pressure head because electrical potential is analogous to hydrostatic pressure. Unless an influx is provided to maintain the pressure head, the reservoir will eventually be drained and the flow will cease, but such an influx is made possible by the use of suitable pumps. See fig. 12.2. The rate of flow of charge (coulomb per second) is defined as the *intensity of the current* (amperes).

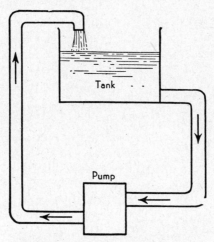

Fig. 12.2. The pressure head of the tank is maintained by the pump, which pumps back into the tank all the water that flows from it.

The Electric Circuit. It should be noted also that in addition to providing the necessary pressure, such pumps must have the ability to pump the required quantity of water if the flow is to be maintained at a constant rate. Electrical batteries are such pumps for electric charges, and obviously the part they play in electrical circuits involves many factors. See fig. 12.3. The study

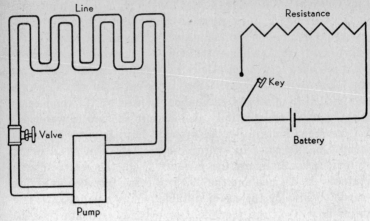

Fig. 12.3. Analogy between a simple series electric circuit containing a battery, a key, and a resistance element and a hydraulic circuit containing a pump, a valve, and a pipe line.

of these factors is the content of this chapter and includes such matters as how batteries accomplish the generation of potential differences, how electric currents are detected, what laws govern their existence, how resistance is offered to the flow of charge, and how electrical devices operate by means of electric currents.

Electromotive Force and Current. There are many kinds of electric pumps, or seats of electromotive force ($E.M.F.$), as they are called. Although the rubbing of an ebonite rod with a piece of fur may develop a very large difference of potential between rod and fur, such a combination does not make a good battery because it is not fast enough; i.e., it cannot maintain a flow of charge of sufficient amount to be of practical importance. So-called chemical batteries, on the other hand, such as the common dry cell and a copper and zinc electrode combination immersed in sulphuric acid (see fig. 12.4), are capable of maintaining a relatively large electrical flow when the terminals are connected externally by means of a copper wire, even though they develop a potential difference of only a volt or so. Chemical disintegration

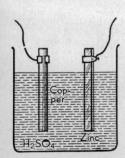

Fig. 12.4. A rod of copper and a bar of zinc immersed in dilute sulphuric acid constitute an electrical battery suitable for many purposes.

of the more active element of the battery (zinc) results in a separation of electric charge. An excess of electrons accumulates on this plate (electrode), while a corresponding deficiency occurs at the other electrode, thus producing a difference in potential between them. Such cells on dead short circuit, i.e., with no appreciable resistance in the circuit, can maintain a flow of many amperes for a few minutes. An *ampere* is a flow of six and a quarter billion billion electrons per second, which is the same as one coulomb per second. *Electric current* is simply the rate of flow of electric charge.

Electric Resistance. Thus it is seen that a battery of some kind is the prime mover in every electric circuit. There is another equally important factor, however, controlling the rate at which charge flows. As the rate of water flow in a pipe depends upon the amount of pressure behind the water, it is also governed by the amount of resistance offered to the flow. Pipes of small cross section, for example, offer greater resistance than do pipes of large cross section, and long pipes offer greater resistance than short ones. Moreover, smooth-walled pipes offer less resistance than rough-walled ones. Analogous factors also enter into electrical considerations. This opposition to electrical flow is what is called *electrical resistance*. It is to be thought of as a physical characteristic of a circuit component just as electromotive force is a specific characteristic of a battery. It has already been pointed out that different substances conduct electricity differently. It is also a fact that the electrical resistance of a wire depends upon its length and cross section.

Ohm's Law. It would be considered a mere truism to state that the flow of fluid through a pipe depends directly upon the push behind it and inversely upon the resistance offered to it. Nevertheless, this is the substance of the most important law of electric currents, namely, *Ohm's Law*. See page 22. Its importance lies to no small extent in its generality. Properly interpreted, it explains practically all current phenomena and gives the physicist the means of calculating otherwise unknown quantities associated with circuits. Specifically, it states that *the current intensity in a circuit* (or a portion of a circuit) *is directly proportional to the electromotive force impressed across the circuit* (or the potential drop across that portion of the circuit), *and is inversely proportional to the resistance of the circuit* (or of the portion of it under consideration). Thus it is seen that if the electromotive force impressed across a given circuit and the resistance of the circuit are known, then the current in the circuit can be readily calculated. Moreover, the potential drop across a resistor can be calculated if the current and the resistance are known. By considering circuits as a whole and then in various parts taken separately, much information can often be determined in terms of very little. This is why Ohm's law is referred to as the single most important law of electric circuits.

Effects of Current. It is appropriate at this point that a question be raised concerning the means used for detecting electric currents if they are merely charges in motion, especially if charges are not capable of direct observation. The answer is that electric currents are recognized by various effects which they have been found to produce. To enumerate briefly, these effects are *chemical, heating,* and *magnetic*. Practically all that is known about electric currents is the result of information obtained indirectly by studying these effects. This again points out how important indirect measurements are in physics.

Chemical Effect of Current. That current is related to chemical activity is not to be doubted in view of the very existence of chemical batteries. In these, the chemical reaction separates charges, thereby causing negative charge to accumulate on one electrode and positive charge on the other. Thus all batteries have positive and negative terminals. When a closed circuit is connected to these terminals, or electrodes, not only is a current established but its direction also is specified.

Direction of Current. It is a matter of convention that the *direction of a current* is taken to be from the positive terminal of the battery to the negative, through the external circuit. This is based upon the idea that unit positive charge is repelled from the positive terminal and is attracted toward the negative terminal via the conducting wire. Of course, the electron theory considers the particles which actually move through solid conductors to be negative electrons, and they pass from the negative electrode where the surplus exists, to the positive electrode which represents the deficiency of electrons. Nevertheless, the conventional direction is still adhered to by most writers, on the grounds that electrical terminology was developed before the electron was ever discovered, and after all it does make sense to speak of the reduction of the deficiency of electrons from the positive to the negative terminal as the electrons flow the other way. It is not so much a question of right versus wrong as it is a matter of self-consistency in describing phenomena incapable of being observed directly.

Electrolysis. Metallic wires are not the only conductors of electricity. Solutions of many chemical salts conduct electricity by the formation of particles called *ions*, electrically charged either positively or negatively, which are attracted or repelled respectively from charged electrodes immersed in them. Such solutions are called *electrolytes*. When current is passed through an electrolyte in a cell the positive ions, called *cations*, are drawn toward the negative terminal, and the negative ions, called *anions*, are drawn toward the positive terminal. These ions get their names from the terms *cathode* and *anode*, the names given to the negative and positive electrodes respectively of the electrolytic cell, the cathode being the terminal where current leaves the cell, and the anode, the one where current enters. Metallic ions, being positive, are deposited out on the cathode of electrolytic cells to which they are drawn. This is the basic principle of electroplating. Silver may thus be electroplated on a spoon, for example, by immersing the spoon in a bath of some silver salt through which an electric current is passed, if the spoon is connected to the negative terminal of the cell. See fig. 12.5. Faraday studied the phenomena of electrolysis and formulated laws governing it. These are considered basic to electrochemical industries even today. Essentially they express a relationship between the mass of material deposited in terms of the total charge passed through the electrolyte (total charge equals current intensity multiplied by time).

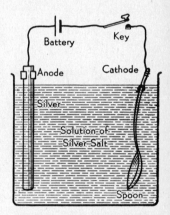

Fig. 12.5. Electroplating. Silver is deposited upon the spoon, which is connected to the cathode of the cell, when the circuit is closed.

The Legal Unit of Current. Electrolysis has also provided the means of specifying the unit of electric current, the *inter-*

national legal ampere, which is defined in terms of the amount of silver deposited per second out of a standard solution of silver nitrate. Thus the study of the chemical effect of the electric current has yielded considerable information regarding electricity in motion, and has provided a way of standardizing, or more accurately establishing, the legal unit of current.

Heating Effect of Current—Joule's Law.

Let us next consider the heating effect of current. It is a well-known fact that a wire carrying an electric current becomes heated. Witness the modern electric range or electric toaster for visible evidence of this phenomenon. The conversion of electrical energy into heat energy is basic to many of the electrical devices which make this an electrical age. The fundamental principle involved is given in the law named after the English physicist Joule, who discovered that *the rate at which heat is developed in a conductor by an electric current depends upon the second power of the current and upon the resistance of the conductor.* Thus if the current is doubled, the rate of heating is quadrupled. Also, coils with large resistance heat more rapidly than coils with low resistance providing the current is the same in each. Because of the entrance of the resistance factor into this picture the conclusion is suggested that Joule's law must be related to Ohm's law. This is found to be true both theoretically and experimentally.

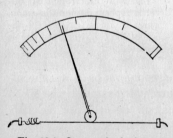

Fig. 12.6. One method of measuring current intensity is with the "hot-wire" ammeter. The wire is wrapped around a spindle which is rotated as the wire expands owing to the heat developed in it by the current. A pointer connected to the spindle swings past a graduated scale.

Electric Power.

As a result of this relationship still another important relationship is found to hold, namely that the product of the potential drop, in volts, across a circuit such as that of an electric heater or electric light, when multiplied by the current, in amperes, through the circuit gives the *power*, in watts, or the rate at which electrical energy is dissipated in heat. This becomes a practical matter when one is concerned with the cost of electrical energy, which varies from approximately one cent to ten cents per kilowatt hour depending upon the locality and the circumstances. A *kilowatt* is one thousand watts, and a *kilowatt hour* represents the amount of energy consumed in one hour at the rate of one kilowatt. Thus at a place where electrical energy cost 10 cents per kwh. (kilowatt hour), a 100-watt electric light bulb can be run for 10 hours at a cost of 10 cents, or one hour for one cent. A flatiron probably costs several times as much to operate as this, and a toaster, about the same as a flatiron.

The heating effect of current is utilized in certain types of *ammeters,* which are the instruments used for measuring current intensity. In the case of one such instrument a wire is wrapped around a spindle upon which is mounted a pointer. See fig. 12.6. As current is established in the wire, heat is developed and the wire lengthens owing to thermal expansion. This causes the spindle to be rotated through a certain angle depending upon the intensity of the current, making the pointer sweep across a portion of a graduated scale placed behind it. Although this principle is readily understood, it is not the one upon which most ammeters actually operate. These depend for their operation upon magnetic effect which will now be described.

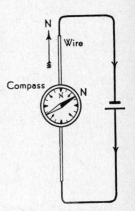

Fig. 12.7. A compass needle placed on top of a horizontal wire carrying current toward the north is deflected eastward. Oersted discovered this effect.

Magnetic Effect of Current. The Danish physicist Oersted was the first to observe that a compass needle is disturbed if it is located in the vicinity of a current-carrying wire. See fig. 12.7. Further examination shows the existence of a magnetic field completely encircling the current-carrying conductor. This is found to be always the case; i.e., no current can be established in a wire unless at the same time a magnetic field is simultaneously set up encircling the wire. See fig. 12.8.

Thus all electric currents have magnetic fields associated with them. It should be noted especially, however, that the current-carrying wire does not itself become magnetized. In fact, the wire is more often than not made of copper, a nonmagnetic substance, and therefore cannot become a magnet. The magnetic

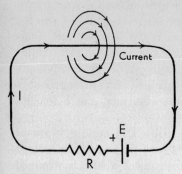

Fig. 12.8. The magnetic field encircles the current-carrying conductor as shown.

effect is produced in the region around the wire since the field encircles it rather than radiates out from it.

The Electromagnet. If a wire is wound into a long solenoidal coil, i.e., a configuration such as would be obtained by winding a single layer of turns on a long cylindrical core and then withdrawing the core, the coil acts like a magnet when current is passed through it, and displays a north and a south magnetic pole. See fig. 12.9. As a matter of fact, a single loop of wire displays a north and a south face; the advantage gained by winding many turns of wire is to increase the intensity of the magnetic field, thus formed, in direct proportion to the number of turns. If a bar of iron is inserted as a core into such a solenoid, the magnetic

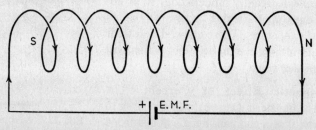

Fig. 12.9. A long solenoid carrying current of intensity I acts like a bar magnet with north and south poles as indicated.

field is also intensified by induction in direct proportion to the permeability of the iron. Such a device is called an electromagnet, and its advantages are obvious. By the mere flick of a switch an electric current can be passed through such an arrangement so as to give it the power to pick up iron objects. By another flick of the switch the circuit can be broken and the electromagnet completely demagnetized, with the result that whatever iron object was clinging to the device is automatically released. Applications of this phenomenon to magnetic cranes, magnetic chucks, and a variety of electromagnetic devices are immediately suggested.

Side Thrust Due to Magnetic Field.

There is one very important electromagnetic phenomenon which alone is responsible for much of the present mechanized age. This is the principle which is basic to the operation of the electric motor. The simple phenomenon is as follows: a wire stretched transversely between the pole pieces of a magnet, either permanent or electric, does not stay put if a current is established in it, but is moved sidewise. Specifically, a current-carrying conductor stretched perpendicularly across a magnetic field is caused to move sidewise, i.e., perpendicularly both with respect to its own direction and with respect to the direction of the magnetic field, by the mutual action of the field of the magnet and the encircling field developed about the conductor. This is because the magnetic field encircling the current-carrying wire interacts with the magnetic field across which the wire is stretched. See fig. 12.10 where the dot symbol indicates that the current is directed out of the paper.

It can be seen that the field due to the current reinforces the north-south field below the wire, whereas above the wire the one field tends to nullify the other one. Consequently the wire is moved from bottom (where the resulting field is strengthened) to top (where the resultant field is weakened) as if the lines of force could be likened to stretched elastic bands. This phenomenon is referred to as *side thrust*. Because side thrust is a mutual force between wire and magnet, the wire will move if the magnet is stationary (as in fig. 12.10), but the magnet will move if it is free and the wire is securely fastened in place (as in fig. 12.7).

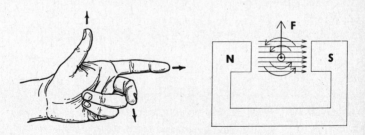

Fig. 12.10. Left-hand motor rule. Forefinger (flux north to south); center finger (current); thumb predicts motion.

Left-Hand Rule. If the forefinger of the left hand is caused to point in the direction of the field of the magnet, i.e., from north to south, and the second finger is made to point in the conventional direction of the current in the wire, i.e., from positive to negative, then the thumb gives the direction of the thrust when these fingers are held so as to be mutually perpendicular in three dimensions. This is called the motor rule. See fig. 12.10.

The Electric Motor. The phenomenon just described represents a transformation of electrical energy into mechanical energy, a matter of considerable practical importance. By careful design an arrangement is readily worked out for converting such a side thrust of a wire into a continuous rotation of a bundle of wires about an axis, thus accounting for the rotation of an electric motor which can be geared or belted to machinery. It is obvious that if a wire carrying a current across a magnetic field is urged sidewise, another wire of the same length, parallel to it but carrying current in the opposite direction, will be urged to move in the opposite direction. Also if these two conductors carry current of the same strength, as they would if joined together in series, forming a rectangular loop, these side thrusts would develop a torque which would tend to rotate the loop to a position across (perpendicular to) the field. (Fig. 12.11.) If, just as this

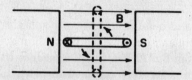

Fig. 12.11. Torque produced by side thrust.

position is assumed, the directions of the current could be reversed, and the reversal could be accomplished at every half revolution, the rectangular loop would rotate continuously, as a motor. Every electromagnetic motor operates on this simple basic principle, from the tiniest flea-power motor to the largest one ever built. In some motors the magnetic part is fixed and the wire conductor arrangement rotates. In others the coils are stationary and the magnetic part rotates. The revolving part is technically referred to as the *armature*.

The D'Arsonval Galvanometer. This consists of a rectangular loop of many turns of wire suspended as an armature vertically between the poles of a permanent magnet (fig. 12.12). This is essentially the same situation as that described in fig. 12.11. When current is established in the coil, a torque is set up which tends to twist the whole suspension against the elastic restoring torque introduced through an external spring. The angle of twist, displayed by a beam of light reflected from a small mirror fastened to the suspension, is proportional to the current in the coil.

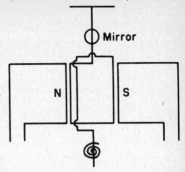

Fig. 12.12. D'Arsonval galvanometer.

Voltmeters and *ammeters* are essentially D'Arsonval galvanometers whose armatures are usually mounted on pivots rather than suspended vertically. Instead of a mirror and light beam playing on a scale to measure the current in the coil, a pointer is fastened to the coil and is made to sweep across a scale mounted in back of it as the coil is twisted.

Electromagnetic Induction. And now for the last phenomenon to be discussed in this description of electric currents, the phenomenon of electromagnetic induction. Again it was Faraday who, over 100 years ago, discovered the inverse of the side-thrust phenomenon described in the preceding paragraphs. Whereas in the former case a wire is moved across a magnetic field merely by the establishment of a current in it, Faraday discovered that the result of mechanically pushing a wire, which forms part of a closed electrical circuit, across a magnetic field is the establishment of current in the circuit as if by a battery. In other words he discovered a way of inducing an electric current in a circuit merely by moving something. In still other words he discovered a way of generating the electromotive force necessary to produce electric current, without the aid of batteries. This simple discovery paved the way for the modern generators which, operated either by steam or water power, are the very mainstays of today's electrical civilization. Without the generators there is no electrical power. And basic to all generators is the simple matter of electric wires being pushed across magnetic fields by the rotation of armatures. The generator principle is simply the inverse of the motor principle. It represents a transformation of mechanical energy into electrical energy.

Lenz's Law. It should be mentioned that the process described above is but one way of inducing an electric current in a circuit. Further study shows that the situation is far more general than has just been described. Any change whatever in the *status quo* of an electric circuit which causes a variation in the magnetic characteristics of the circuit, is sufficient to establish momentary currents in neighboring circuits. This is the gist of a law of induced currents known as *Lenz's Law*. The induction associated with a changing current is basic to the operation of the electrical transformer and other so-called alternating-current devices. The study of these devices, however, including the spark coil, the telephone and telegraph, the radio, etc., would carry us far beyond the descriptive stage and so must be postponed for a more exhaustive consideration by those students who are willing and able to develop a more mathematical language. Suffice it to say that the principles discussed above are basic to the operation of these devices and account perfectly for their behavior.

Summary. In this chapter the various effects of electric currents have been described. It has been shown that the study of these effects has developed considerable information about electric charges in motion, although the charges themselves are not capable of being directly observed. This information has already been carefully organized into a body of logically developed laws and principles during the last one hundred years or so, and today we live in an electrical age because of the many practical applications of these principles. Of these the principles of the motor and the generator are especially important.

QUESTIONS

1. What is the nature of the electric current?
2. Explain the function of the battery in the electric circuit.
3. Using a hydraulic analogy discuss the meaning of Ohm's law.
4. Why is an electrostatic machine not suitable as a battery?
5. Describe the process of electroplating.
6. List three major effects of the electric current.
7. Distinguish between the "direction of current" and the direction of electron flow.

8. Which costs n.ore to operate, a 100-watt bulb for 10 hours or a 10-watt bulb for 150 hours?
9. What was Oersted's discovery?
10. Explain how the electromagnet functions.
11. What is the left-hand rule?
12. How is the international ampere defined?

CHAPTER XIII

ELECTRONIC CONSIDERATIONS ATOMIC AND NUCLEAR PHENOMENA

INTRODUCTION TO MODERN PHYSICS, ELECTRICAL DISCHARGES IN GASES, X RAYS, ELECTRONICS, RADIOACTIVITY, NUCLEAR AND SOLID STATE PHYSICS

The Situation about 1890. Shortly before the dawn of the twentieth century the subject of physics was in a rather unique position with reference to the other sciences and the many fields of learning. This was about the end of the so-called classical period, which was characterized by the Newtonian point of view. The situation was unique because the physicist, in his attempt to explain physical phenomena, was apparently almost completely successful. Practically all that had been observed was understandable in terms of simple concepts and laws such as Newton's laws of motion, etc. These were so carefully and so logically developed that the physicist seemed to have the situation completely in hand. In a fashion little short of miraculous, new observations had been fitted into places apparently made for them in the theories and hypotheses extending over a hundred or more years. On the theoretical side it was felt that physics was fast approaching the status of a dead classical subject so completely worked out in detail as to leave very little of interest, if anything at all, for future generations to unravel. At this time famous physicists made statements to the effect that all the important discoveries in physics had been made and that most of the romance in the subject was gone; all that was left for future generations was the somewhat colorless task of extending the precision of the measurements of known physical constants and relationships.

That this view was all wrong is the story of modern physics, a story that has completely revolutionized the whole outlook of physical science. That the physicist has been able to appreciate the change in the situation is even more significant. In earlier chapters references have been made to the new physics, meaning the developments since 1900 or thereabouts. The purpose of this chapter is the consideration of some of the ideas and concepts of modern physics as they have been developed since the discovery of the electron shortly before that.

Electrical Discharges in Gases. About 1895 the eminent English physicst Sir. J. J. Thomson, following earlier work by Crookes, was experimenting with the conduction of electricity through gases. Among other things he observed that dry air at atmospheric pressure is a rather good insulator, requiring some thirty thousand volts per centimeter to break it down, i.e., to make it pass charges of electricity in the form of the electric spark. It was found, however, that when the pressure of the air was reduced, as by a vacuum pump for example, the necessary voltage for breakdown was appreciably diminished, but not without a very noticeable change in the character of the discharge. These experiments were first carried out at a time when vacuum pumps were not only scarce but exceedingly inefficient. Today, however, because of technical advances, vacuum pumps are available which make it possible to perform these experiments in a demonstration lecture right before the very eyes of students.

Discharge Tube Demonstration. If a long, slender glass tube, say about two feet long and an inch or two in diameter, with electrodes sealed in at its closed ends, is connected to a modern vacuum pump through a tube joined to it at one side, a succession of interesting phenomena will be observed as the air is very gradually withdrawn from the tube while a large potential difference is maintained across the electrodes by a spark coil or some other source of several thousand volts. See fig. 13.1. In a darkened room the experiment is enhanced. At atmospheric pressure, or at the start of the experiment, no discharge whatever is visible between the electrodes; but very soon after the vacuum pump starts a long, ragged, sparklike discharge occurs. As the pressure is further reduced the sparklike discharge gives way to irregular violet streaks of light. Soon thereafter the whole tube becomes lighted by a purplish glow. At this state the pressure is probably not more than a few millimeters of mercury, or somewhat more than a few thousandths of atmospheric pressure.

Presently, the general glow turns somewhat pinkish and then becomes quite pink. In the meantime, a very definite blue glow appears near the negative electrode. This is called the *negative glow*. Between this glow and the main body of the pink discharge

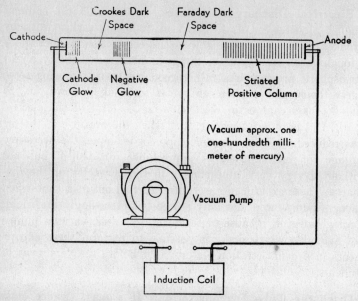

Fig. 13.1. Discharge tube phenomena.

a relatively dark space develops. This is called the *Faraday dark space*. The pink glow, which gradually shortens to make room for the Faraday dark space, is known as the *positive glow* or the *positive column*. Soon this positive column presents a very unique appearance by breaking up into innumerable striations while at the same time the pink color fades into a pale orange color, even turning white. The shortening continues as the Faraday dark space and the negative glow seem to break away from the cathode and to move nearer the positive electrode, leaving a second dark space between the negative glow and the cathode itself. This dark space has become known as the *Crookes dark space*. The cathode itself also glows, with a violet light known as the *cathode glow*.

By this time the pressure has been reduced to probably a matter of tenths or even hundredths of a millimeter of mercury. As the exhaustion continues the striations of the positive column present a more and more fluted appearance, becoming coarser and coarser. The positive column, which is now definitely associated with the positive end of the tube, becomes shorter and shorter. The Faraday dark space and negative glow move farther and farther away from the cathode, with the result that the Crookes dark space gets larger and larger, crowding all else right out of the positive end of the tube. By the time the pressure is reduced to about a thousandth of a millimeter of mercury the whole positive column has been thus completely crowded out and the Crookes dark space completely fills the whole tube. That is to say, the whole tube appears very dark except that the glass walls glow with a peculiar green light called *fluorescence*. Fluorescence is the name given to the process by which radiation of a given wave length is transformed into a radiation of a different wave length (usually longer). In this case fluorescence is produced by cathode-ray bombardment from the negative electrode.

Cathode Rays. At this stage of exhaustion it is noted that any object inside the tube between the electrodes casts a shadow on the positive end of the tube. This indicates the existence of some sort of radiation from the cathode toward the anode. In addition to producing fluorescence, the radiations travel in straight lines, forming sharp shadows of objects in their paths. It is also found that the radiation can be focused by using a spherically concave cathode. When it is thus focused upon a target, heat is generated at the spot on which the rays fall. Because the radiation emanates from the cathode the name of *cathode rays* has been given to it. It was shown soon after their discovery that these cathode rays can be deflected by magnetic and electric fields. This proves that they consist of electrically charged particles constituting an electric current. The fact that the deflection in a magnetic field is a side thrust phenomenon directed oppositely to that predicted by the left-hand side thrust rule, unless the conventional direction of current is assumed rather than the direction of the particle flow from the cathode, proves that the current consists of negatively charged particles (now called *electrons*) which stream from regions of low potential to regions of high potential. These are the same electrons to which earlier references have been made in this text in connection with the study of matter and its electrical properties. It was the discovery of these electrons by considerations of electrical discharges in gases that opened up the so-called new era in physics.

X Rays. Soon after the discovery of cathode rays it was observed by Roentgen that when they were focused upon a metallic target inside the same vacuum tube, a secondary radiation was emitted. This is the so-called *X-ray radiation* and is not to be confused with the cathode radiation. See fig. 13.2. It is of the wave variety like light and is characterized particularly by its ability to penetrate relatively opaque substances. It not only makes possible the modern X-ray shadow photographs but is used extensively today in hospitals for therapeutic work. The nature of the X radiation depends upon several factors, includ-

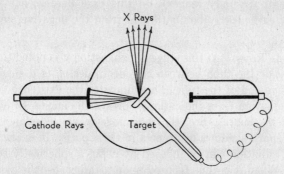

Fig. 13.2. X rays may be produced by cathode rays bombarding a target.

ing the potential drop across the tube, the degree of vacuum, and the substance of which the target is made. X rays are sometimes called Roentgen rays after their discoverer.

Applications of Discharge Tubes. Partially exhausted tubes through which an electrical discharge is produced, as in the experiment described above, are called *Geissler tubes*. The color of the discharge depends upon the gas used. The modern neon advertising signs are nothing but such discharge tubes, using the gas neon, however, instead of air because of the characteristic red color of neon when activated in this manner.

Modern fluorescent lighting, both office and household, represents another application of discharge tubes. Instead of utilizing the incandescent filament as a source of light, the phenomenon of fluorescence is utilized by coating the walls of the tube with various chemicals capable of producing different colors. Such lamps are more economical to operate than filament-type lamps.

Quantitative Aspects of the Electron. J. J. Thomson not only established the fact that cathode rays were streams of negatively charged electrons; he also measured in a very ingenious way a specific characteristic of the electron, namely the ratio of its charge to its mass. A stream of electrons was first subjected to a transverse electric field of measured intensity, as a consequence of which the stream was deflected into a circular path by electrostatic attraction. A measurement of the radius of curvature of this path was made. Then a magnetic field of measured intensity was applied perpendicularly to the electron stream and also perpendicularly to the electric field, and so directed as to produce by magnetic side thrust a counterdeflection of the stream just nullifying the effect of the electric field. The forces which are here balanced are centripetal forces (see p. 42) which can be evaluated in terms of the mass of the electron, the charge on the electron, the radius of curvature of the path when only one field is applied, the intensity of the electric field, and the intensity of the magnetic field, all of which are directly measurable except for the first two. The ratio of charge to mass (e/m) is found to be 1.76×10^{11} coulombs per kilogram (176 billion coulombs per kilogram). This is the same as 528 trillion stat-coulombs per gram. Thus we see that although the electron is so small that it can be seen only by the eye of the mind it is nevertheless possible for us to actually measure a physical characteristic of it.

Millikan's Determination of the Charge on the Electron. On p. 134 reference was made to the fact that around 1910 the American physicist Millikan succeeded in measuring the charge (e) on the electron. His oil-drop experiment is now recognized as one of the famous experiments of all time. Fine droplets of oil are sprayed into a region between two metal plates which can be maintained at a controlled difference of potential. A particular droplet is watched through a microscope and its rate of fall under the influence of gravity alone is first determined. Then the electric field is established between the plates and it is observed that electric charges are picked up by individual droplets, making themselves evident by abrupt changes in the droplet's motion. In terms of measurable quantities such as the density and viscosity of the oil, the velocity of the droplet, etc., it becomes possible to determine, in a relative manner, the amount of charge acquired by a given droplet. The interesting fact is that every charge observed is a multiple of a certain fundamental charge; and assuming that a droplet of oil cannot acquire less than a single electronic charge, it is concluded that this observed fundamental charge is the charge on an electron. It amounts to 4.80×10^{-10} E.S.U. (stat-coulombs) or 1.6×10^{-19} coulombs, and is negative.

Using this value for e and J. J. Thomson's value for e/m (1.76×10^{11} coulombs per kilogram), it follows that m (the mass of an electron) is 9×10^{-28} grams.

These two experiments are singled out for mention here not only because they are now considered classical experiments in modern physics but also because they exemplify how the "new" physics is extrapolated from the "old" physics. In the first one it was necessary to appreciate the classical concepts of mass, velocity, and centripetal force, as well as the concepts of electric and magnetic field, electrostatic attraction, electromagnetic side thrust, etc., to give quantitative meaning to the newer concept of the electron. In the second one it was necessary to appreciate concepts associated with falling bodies, density, viscosity, etc. in order to make sense out of the measurements. Thus it follows that a real appreciation of modern physics requires a substantial background in classical physics.

Thermionic Emission. With the discovery of the electron came the understanding of another phenomenon which is fundamental to modern communication, i.e., the *Edison effect,* or the thermionic emission of electrons. Heated filaments in vacuum tubes are found to emit electrons which can be drawn across an empty region by electrostatic attraction toward positively charged plates. See fig. 13.3. Such electronic behavior is the basic principle of all radio tubes and of the host of so-called control tubes for innumerable electronic appliances. See fig. 13.4. The modern X-ray tube utilizes thermionic emission for the production of

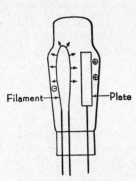

Fig. 13.3. When a filament is heated in vacuum, electrons are emitted by it. These electrons are readily attracted to a positively charged plate, producing a current between filament and plate. This explains the Edison effect of a current between plate and filament when the former is charged positively.

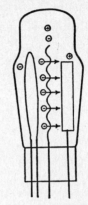

Fig. 13.4. The rectifying action of the grid in a three-electrode vacuum tube. The grid, if alternately charged positively and negatively, acts so as to control the flow of electrons from filament to plate. It allows the flow to take place only when it is positively charged.

electrons. Once obtained by emission from a hot filament, the electrons are directed toward a target from which X rays are emitted.

Photoelectric Effect. Another interesting electronic phenomenon is that of the emission of electrons from the surfaces of certain substances by the action of light or radiation generally, including X rays. This is called the *photoelectric effect*. It is basic to the operation of the photoelectric cell, or electric eye as it is more popularly known. When light shines upon these cells electric currents are established, but when the light is shut off the current is interrupted. See fig. 13.5. Thus an electric current can be controlled just by regulating the amount of light falling upon the cell. Because of this phenomenon the photoelectric cell is a

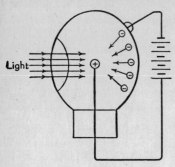

Fig. 13.5. The action of a photoelectric cell. Light falling upon the inside of a vacuum tube coated with certain active substances causes electrons to be emitted. These electrons can be attracted to a positively charged electrode in the tube to produce an electric current capable of being controlled by variations in light intensity.

very useful device for controlling electric currents. It plays a very important part in the operation of sound movies and in television. In the case of the former, a beam of light is varied in intensity or direction by certain markings on the side of the film, whereupon an electric current is varied accordingly to control the sound mechanism. In the case of television, light reflected from the subject is picked up by a scanning device in which an electric current is varied according to the variations in intensity of the reflected light. In the television receiving set the operations are essentially reversed and a varying current is translated into a visual image which is cast upon a screen.

The explanation of the photoelectric effect utilizes the quantum theory of Planck. Although this theory was developed to explain the nature of heat radiation (p. 123), Einstein suggested its use to explain two aspects of the photoelectric effect, (1) that the emission of electrons takes place only when the frequency of the incident light exceeds a certain critical value known as the threshold frequency, and (2) that the velocity, and therefore the kinetic energy, of the emitted electrons depends only on the frequency of the light and not upon the intensity of the light as might be expected. This proposal was later verified experimentally by Millikan, thus increasing the physicist's confidence in the quantum theory.

Electronics. The general term "electronics" is given to all phenomena dealing with the electron. Applications of our knowledge in this area of physics are numerous. Since many of them lie in the field of communications, such as radio, radar, television, etc. some people erroneously conclude that electronics means simply these things. Today, however, industrial and medical applications are widespread. X rays actually come under the heading of electronics. Indeed the field of electronics is practically limitless since all matter is made of electrons. The trick is to liberate them by such methods as thermionic emission and photoelectric emission, or by literally dragging them out of matter by the application of high voltages under vacuum conditions. Once they are liberated they can be manipulated in endless ways by electric fields, magnetic fields, etc. to serve man's will. It is obvious why this age is often referred to as the electronic age and why modern physics is so involved with electronics.

Radioactivity. At about the same time that Roentgen discovered X rays (1905–1906), Becquerel made a discovery which was to prove fully as important as the isloation of the electron or the discovery of X rays. He observed that certain uranium salts spontaneously emit radiations which can be studied by the effects they produce when subjected to electric and magnetic fields. Somewhat later Madame and Pierre Curie concentrated from uranium ores certain substances which were named polonium and radium and which displayed the phenomenon of spontaneous disintegration to a very marked degree. Today the phenomenon is called *radioactivity*. Radioactive substances emit electrically charged particles along with radiation similar to X rays. The radiations were named alpha, beta, and gamma before their natures were understood. The *alpha* radiation consists of particles now known to be the nuclei of helium atoms. Since they carry two extra positive charges they are called doubly ionized helium atoms. The *beta* radiations are streams of negative electrons, and the *gamma* radiation is wavelike rather than corpuscular, being a very short-wave type of X ray. See fig. 13.6.

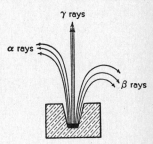

Fig. 13.6. Radioactive substance in a slot in a lead block. The action of a magnetic field transverse to the diagram upon α (alpha) rays, β (beta) rays, and γ (gamma) rays respectively. Beta rays are sharply curved one way; alpha rays are less sharply curved the other way; gamma rays are unaffected.

Atomic Structure. The discovery of radioactive disintegration obviously suggested that the atom of matter, instead of being an indivisible entity as had been supposed up until that time, in reality had a structure of its own involving subatomic components. Modern physics has been concerned largely with the nature of this structure and of its component parts. Alpha particles have been used as bullets to bombard atoms in attempts to learn more about this structure. As indicated earlier (p. 88), the present view of the atom is that of a positively charged nucleus enveloped or surrounded by some kind of distribution of negative electric charge. It has been found possible even to make substances artificially radioactive by bombardment with atomic particles accelerated to high velocities by super-high-voltage devices. Of these the Van de Graaf high-voltage generator and the Lawrence cyclotron have been widely publicized. Both of these devices are capable of endowing particles such as electrons, protons, and other ions with energies equivalent to millions of electron volts. The *electron volt* is the amount of energy acquired by an electron accelerated through a difference of potential of one volt. These devices have really been developed for the purpose of smashing atoms and atomic nuclei to bits in order that the products of such disintegrations may be studied. It is the hope of today that the disintegration of the nucleus may open up new fields of knowledge as did the breaking up of atoms several generations ago.

Positive Rays and Isotopes. Along with the discovery of cathode, or negative, rays streaming from cathode to anode, a positive radiation was also observed. This positive radiation was discovered by the use of perforated cathodes which allowed particles originating at the anode to pass through the perforations in the cathode and be detected at the glass wall opposite to the anode. See fig. 13.7.

Whereas the cathode rays consist of negatively charged particles, all exactly alike, as shown by the action of a magnetic field upon them, the positive rays are found to consist of positively charged ions of various masses and charges. It appears then that atoms of a given chemical element can have different masses, i.e., they can have different physical properties while retaining

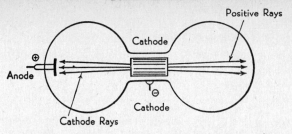

Fig. 13.7. Positive rays are emitted by the anode as well as negative rays by the cathode. These are positive ions. Both sets of rays are capable of deflection by electric and also magnetic fields.

the same chemical properties. These different forms of a given element are called *isotopes* of that element. The least massive of all isotopes is one which has a mass approximately equal to that of a hydrogen atom. It is called a *proton*. Its charge is numerically equal to that of the electron, but the sign of the charge is positive rather than negative. Its mass however, is some **1845** times as great as the mass of the electron. Thus it appears that the ordinary hydrogen atom is some kind of combination of a proton and an electron. Hydrogen also has an isotope with twice the mass of the proton but with the same electrical charge. This is called the *deuteron*. Whereas two atoms of ordinary hydrogen combine chemically with one atom of oxygen to form a molecule of ordinary water, such a combination using deuterons instead of protons gives what is called *heavy water,* whose physical properties are somewhat different from those of ordinary water. Practically all the elements have one or more isotopes.

Bohr Theory. These elementary facts regarding electrical particles led to the development of a succession of theories pertaining to atomic structure. Probably the most famous of these is the Bohr theory of 1913 previously mentioned in this text. On this theory the hydrogen atom consists of a single proton about which a single electron revolves like the moon around the earth. See fig. 13.8. The atom of helium has a positively charged nucleus about which two electrons revolve, but the nucleus has a mass four times that of the proton. In this way all the elementary atoms were thought to be made up of combinations of positive and negative charges of electricity forming dynamical systems of varying complexity. The original success of this theory was its explanation of optical spectra, about which more will follow in succeeding chapters. In many details, however, this theory has been found inadequate and is no longer accepted as final. Yet the views which have supplanted this theory are not altogether suitable for pictorial representation, and so the Bohr theory is still referred to in many popular discussions of atomic structure.

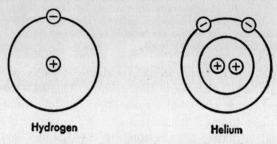

Hydrogen　　　　　Helium

Fig. 13.8. Bohr's concept of the hydrogen and the helium atom, with negative electricity circulating around positively charged nuclei. The helium nucleus is the alpha particle of radioactivity, and has four times the mass of the proton.

The *Bohr model* pictures the outer structure of the atom as consisting of electrons which revolve in elliptical orbits about the nucleus, a single electron in the case of hydrogen, and an increasing number as more complicated atoms are considered up to lawrencium, the most complicated one in the periodic table of the natural elements. In the Bohr model various orbits are possible for the electrons depending upon the excitation state of the atom, the larger ones being associated with more energy than the smaller ones. Discrete *energy levels* appear to be more important, however, than the concept of orbits. Optical and X-ray spectra are explained by the emission of radiation as electrons fall in from outer to inner orbits or from higher to lower energy levels. The spectrum of hydrogen is practically completely so accounted for, but discrepancies are found in spectra of more complex substances.

The differences in energy levels between the Bohr orbits, and consequently the amounts of radiation emitted, are multiples of the basic *quantum* of action suggested by Planck's quantum theory (see p. 123). Thus it is supposed that atomic phenomena are governed by the laws of *quantum mechanics* rather than those of Newtonian or classical mechanics. According to the *uncertainty principle* of quantum mechanics, there is an ultimate limit to the precision with which events can be determined. There are basic uncertainties. For example, the position and the momentum of a particle cannot both be measured simultaneously with unlimited precision. As the precision of determination of the one increases, the limit of certainty of the other decreases. The product of the uncertainties of these two quantities is approximately equal to Planck's constant h, the fundamental constant appearing in Planck's quantum theory. This principle becomes significant only for particles of atomic size or smaller. It is completely obscured in the case of phenomena in the macroscopic world.

For many years after the discovery of the electron and the proton, these two particles were considered the sole constituents of matter. It was not until the fourth decade of the present century that this number was increased by the discovery of the neutron and then within a few months by the discovery of the positive electron. The *neutron* is a particle carrying no net charge and having a mass approximately equal to that of the proton. The *positive electron* (or *positron*) carries a positive charge but otherwise has all the characteristics of the negative electron. It now appears that the nuclei of all atoms are composed fundamentally of protons and neutrons. Much of today's research work is devoted to the possible discovery of additional fundamental particles and to the study of atomic nuclei. Only yesterday the atom was found to have a structure, and today the attention of physicists is focused upon the question of the structure of one of its component parts, namely the nucleus. Tomorrow may bring yet more surprising discoveries.

The Nucleus of the Atom. The nucleus of the atom appears to be made up of components known as nucleons which are held together by forces not well understood. Consequently the details of its structure are not yet altogether clear. Two basically different theories are current, the liquid drop model and the shell model. Nuclear forces considerably greater in magnitude than coulomb forces of electrostatic attraction or repulsion are considered in the first model to behave in a manner analogous to surface tension forces to cause the components of the nucleus to coalesce into something analogous to a liquid drop. In the other model closed shells involving discrete energy levels are postulated after the manner of the shells or energy levels associated with atomic structure.

Nuclear components include the *proton*, the *neutron*, the *positron*, various *mesons* (particles considerably more massive than electrons and charged negatively or positively), and innumerable other nondescript particles called *pions*, *muons*, *strange particles*, and so-called *anti-particles* of various kinds.

Devices for detecting nuclear components include the following. The *Geiger-Mueller counter* is a device in which a wire mounted coaxially inside a metal cylinder, but insulated electrically from it and raised to a high potential difference with respect to it, discharges momentarily when a charged particle (alpha or beta) happens to pass through a very thin metal foil window into the tube. The *Wilson cloud chamber* is a device in which droplets of water vapor condensing upon minute particles when the vapor is expanded form tracks which can be observed visually or photographically, and which indicate the recent presence of said particles. *Photographic emulsions* can be used to detect particles. *Bubble chambers* use supersaturated liquid hydrogen to make tracks of particles. *Scintillation counters* consist of multiple photo cells so oriented as to multiply minute photoelectric currents.

Cosmic Rays. Cosmic rays constitute a source of nuclear particles. Originating in outer space, they have been discovered to penetrate the earth's atmosphere. Although primarily protons, they produce secondary cosmic radiation in shower bursts as they collide with matter.

Nuclear Reactions and Transformations. Rutherford in 1919 showed that nitrogen bombarded by alpha particles (helium nuclei) was transformed into oxygen with the release of protons according to the reaction

$$_2He^4 + {}_7N^{14} \rightarrow {}_8O^{17} + {}_1H^1.$$

Also lithium, when bombarded with protons, is transformed as follows to form helium.

$$_1H^1 + {_3}Li^7 \rightarrow {_2}He^4 + {_2}He^4$$

Note. The subscript to the left in the above notation indicates the so-called *atomic number* Z, while the superscript to the right indicates the so-called *atomic mass* A. A is 16 for oxygen, i.e., it is the nearest whole number to the atomic weight. Z is the number of protons in the nucleus, i.e., it specifies the amount of positive charge which the nucleus exhibits. Neutrons were first produced in 1930 by Chadwick, who bombarded beryllium with alpha particles according to the following:

$$_2He^4 + {_4}Be^9 \rightarrow {_6}C^{12} + {_0}N^1$$

Radioactive Decay and Half Life. The spontaneous emission of alpha and beta particles from the naturally (in contrast to the artificially) radioactive substances is accompanied by changes in the position of a substance in the periodic table of the elements. The ejection of an alpha particle represents a loss of two protons and two neutrons, i.e., the mass number A *decreases* by four units whereas the atomic number Z *decreases* by two. The emission of a beta particle means a loss of negligible mass but an *increase* of the atomic number Z by one unit. Thus the radioactive materials decay, usually by well-defined steps, to some final stable substance, usually some isotope of lead for the heavier elements like radium, uranium, thorium, actinium, etc. The *half life* is that time required for a substance to lose one half its activity. Half lives vary from fractions of a second to millions of years for different isotopes, but each has its own characteristic value.

Nuclear Binding Energy. When nuclear transformations take place and nucleons combine to form nuclei (as when neutrons and protons combine to form helium nuclei), or when nuclei disintegrate naturally (natural radioactivity) or artificially by bombardment with other nucleons (artificial radioactivity), the mass of the nucleus is usually not the same as the combined masses of the component nucleons. The difference is called the *binding energy* in view of Einstein's relationship between mass and energy $E = mc^2$, where c is the velocity of light. The binding energy per nucleon is a function of the mass number A, and is greatest for the elements near the center of the periodic table and least for elements at the extreme ends of the table.

Fission. Early in 1939 the remarkable discovery was made that when slow neutrons are allowed to bombard certain substances, notably the 235 isotope of uranium, nuclei of the latter, under appropriate circumstances, are observed to split into components which have been identified to include atoms of barium and krypton, i.e., substances located near the middle of the periodic table rather than at either end. The process is now referred to as *nuclear fission*. Along with the splitting is observed the liberation of high intensity gamma radiation (electromagnetic waves) and additional neutrons, which themselves can be made to collide with more nuclei of the uranium isotope to produce more components and more radiation and more neutrons, etc. This process can be controlled by the use of moderators like cadmium or water (preferably heavy water), or can be allowed to get out of control. The radiation is the equivalent of the binding energy being released in accordance with the Einstein relation $E = mc^2$, which collectively for the billions and billions of atoms involved in a few pounds of uranium represents a fantastic amount of energy (comparable with the energy released by the explosion of tons of T.N.T.).

The Atomic Bomb. The military possibilities of atomic energy were immediately recognized. This led to the development of the atomic bomb as the war's greatest secret enterprise at a cost reputed to be around two billion dollars. When a critical amount of U235 is accumulated, a self-sustaining nuclear chain reaction is a necessary consequence. This takes place extremely rapidly and with the evolution of a tremendous amount of energy. As a neutron splits the uranium atom, additional neutrons are evolved which split additional atoms, and so on indefinitely until all the U235 is used up. Another important feature of the process is the creation of a new element, plutonium, which also exhibits fission. By methods which, even at the time of this writing, are still carefully guarded military secrets, devastating atomic bombs utilizing nuclear fission were created in the summer of 1945. This marked the first occasion on which man had released sizable amounts of atomic energy, and no one can predict the social and economic consequences of this important event. It took World War II to provide the stimulus for crowding into just a few years an amount of research in nuclear physics that would normally have taken probably scores of years.

Fusion and the Hydrogen Bomb (Thermonuclear Device). Temperatures produced in the fission process are sufficiently great to produce the *fusion* of hydrogen nuclei (protons) and neutrons into helium nuclei (alpha particles) with the release, in the form of radiation, of the excess energy beyond that necessary to bind these nucleons together (hydrogen and helium are at the lower extreme of the periodic table where the binding energies are not the greatest). Thus the *hydrogen bomb,* theoretically unlimited in size, became available following the development of the atom bomb, before which the necessary temperatures were not only unavailable but inconceivable of attainment. Incidentally less radioactive debris is produced by the H bomb than the A bomb.

Molecular Bonding. Molecules, being aggregates of atoms, are held together by various forces. For those molecules which are simple combinations of positive and negative ions, coulomb forces seem to suffice. For diatomic molecules like H^2 the atoms appear to share the two electrons by what are called *exchange forces*. The *Pauli exclusion principle* suggests that when one atom approaches another to form a molecule the electrons of the one penetrate the occupied energy shells of the other, but some are forced into higher states because crowding of shells is not allowed. This explains how attractive forces are developed to hold the molecule together in spite of the coulombic repulsion of two positive nuclei.

Solid State Physics. The atoms of a solid are held together by forces which result in a crystalline structure. In metals the electrons are free to migrate through the crystal from atom to atom because they do not appear to be permanently associated with any particular one.

Whereas classically the vibrational energy of molecules and atoms is assumed to be zero at absolute zero, quantum mechanics predicts a residue of atomic energy at 0° on the Absolute temperature scale. The quantizing of vibrational waves in crystal lattices gives rise to the concept of the *phonon,* an elastic counterpart of the electromagnetic *photon.*

In a solid the energy levels of the component atoms become broadened to form what are called *energy bands*. At absolute zero (see p. 104) the electrons are in the lowest possible band or level, i.e., the lowest levels are filled, but as the temperature is raised electrons become excited and tend to occupy higher bands. In metals the so-called conduction bands are not completely filled. *Conductors, semiconductors,* and *insulators* are distinguished by the extent to which the conduction (upper) bands are filled. The so-called *Fermi level* in a substance is that level at a given temperature for which the probability that an electron occupies any available state at that energy is 50%.

Impurity semiconductors are formed when small amounts of a foreign substance are added (an example is arsenic added to a germanium crystal). This alters the distribution of electrons in the valence band and conduction band (separated by the Fermi level). An arsenic atom provides an extra conduction electron and is called a *donor* impurity and produces a so-called n-type semiconductor. *Acceptor* impurities result in so-called p-type semiconductors in which missing electrons produce so-called *holes* available for conduction.

The *transistor* is an n-p-n junction which has found important applications as an amplifier in a device which has replaced the electronic vacuum tube in many electronic circuits.

The *laser* is a solid-state device in which atoms are excited from lower to higher energy states by a novel process in which light of higher frequency than that emitted by the atoms when they return to their original states is so-to-speak pumped into them. This results in a greater than normal number of atoms in the higher states so that when radiation is stimulated, it is greatly amplified while the frequency remains the same as that of the stimulating source and therefore is said to be coherent.

Summary. The subject of modern physics has been developed in this chapter from the early experiments in electrical conduction through gases to the modern super-high-voltage machines used in nuclear disintegration experiments. Although many topics have been mentioned here, it must not be assumed by the student that the discussion is anything like complete. Modern physics is far too large a subject to be treated exhaustively in a single chapter, but for purposes of this descriptive survey of physics the treatment should be adequate to give the beginning student a proper orientation. On the other hand, this chapter has been more detailed than previous ones because of the importance to modern society of what is called modern physics. An attempt has been made to acquaint the student with the relatively large number of technical terms which have crept into the vocabulary in recent years. These include the names of the many atomic particles; the notation used to describe nuclear transformations; the various concepts of so-called solid state physics and such practical devices as the transistor and the laser, just to recall a few. An attempt has also been made to emphasize the need for an understanding of the concepts of so-called classical physics before a real appreciation can be had of those concepts which are unique to modern physics.

Much of modern physics also comes under the heading of optics and as such will be considered in succeeding chapters. We turn next to a description of optical phenomena and the question of the nature of light.

QUESTIONS

1. Describe briefly the appearance of the electric discharge in an elongated glass tube as the pressure of the air within is gradually reduced.
2. Distinguish between cathode rays and X rays.
3. What is meant by the photoelectric effect?
4. To what, in general, does the term "electronics" refer?
5. What is meant by radioactivity?
6. Distinguish between alpha particles, beta particles, and gamma rays.
7. What is the Bohr model of the hydrogen atom?
8. What are isotopes?
9. How does a neutron differ from an electron or a proton?
10. What is meant by nuclear fission?

CHAPTER XIV

OPTICAL CONSIDERATIONS

PHOTOMETRY, LAWS OF GEOMETRIC OPTICS, OPTICAL INSTRUMENTS

Scope of Optics. Of all physical phenomena there are probably none more fascinating and of more immediate interest to the average person than those related to human vision. The worlds of color, sunshine, geometric design, and fine instruments also contribute to the complex realm of optics. This field also includes such studies as illumination, photography, microscopy, spectroscopy, optometry, and polarization, as well as the study of certain philosophical questions such as the nature of light itself. Thus it is seen that optics is indeed a very broad subject; yet at the same time, it is but a part of that larger subject, physics.

Complex Nature of This Study. From a strictly logical point of view, this subdivision of physics is less well adapted to the methods employed in the foregoing parts of this text, i.e., the logical development from one topic to another, than those which have preceded it. This is not only because of the complexity of the subject but also because of the lack of any logical starting point. On the historical side, many complex and apparently unrelated facts were known about optics before other simpler ones were discovered. On the philosophical side, too, there has existed the perpetual controversy between the proponents of the corpuscular theory and those favoring the wave theory of the nature of light. Moreover, the points of contact between this and other parts of physics are numerous; yet at the same time they are as diversified as the studies of geometry and the colors of the rainbow. Indeed it is this diversity that provides us with a key to a systematic study of optical phenomena.

Geometric vs. Physical Optics. The study naturally divides itself into two major divisions which are designated as *geometric optics* and *physical optics*. The former deals with the light rays traveling in straight lines and the images which are produced by reflection and refraction, whereas the latter deals with the physical nature of light and related phenomena. We shall consider each of these in turn, starting with geometric optics partly because of the historical development of the subject. The more recent researches of the last decade or so have had a particularly important bearing upon physical optics. This chapter will deal with geometric optics, leaving physical optics for the next and final chapter of this book.

Early Theories. It was held by the ancients that light and vision were practically the same. They thought that all light originated in the eye, to which contact was made in some mysterious way with observed objects. It seems strange that they were not disturbed by the fact that objects invisible to blind people were visible to others. A later view credited so-called luminous objects with the ability to emit particles of light capable of traveling in straight lines and endowed with the property of exciting the eye of the observer. That the luminous body does the emitting and the eye does the receiving is the modern view, but whether the radiation itself is corpuscular or wavelike is a matter about which physicists have argued throughout the entire history of physics.

Straight-Line Propagation of Light. One fact, however, has become very certain, namely that light travels in straight lines. This is technically called the *rectilinear propagation* of

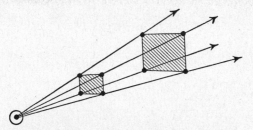

Fig. 14.1. The rectilinear propagation of light.

light, and is proved, in a superficial way at least, by the ability of light to cast well-defined shadows. See fig. 14.1. The pinhole camera also demonstrates this fact. See fig. 14.2. The concept of light rays is thus suggested. Such *rays* either are emitted by self-luminous bodies or are reflected from illuminated bodies to be appreciated by the eye. Objectively speaking, of course, the eye need not be a party to optical phenomena in general.

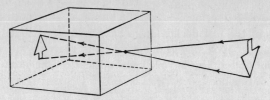

Fig. 14.2. The pinhole camera demonstrates the rectilinear propagation of light.

Photometry. The concept of rays leads to the notion of spherical symmetry of the radiation about a point source of light and suggests a way of measuring amounts of light. This is a field of optics usually referred to as *photometry*. A source of light is compared with a candle, and sources are rated for intensity in units called *candle power*. Thus a 100-candle-power source is approximately 100 times as intense as a specified type of candle. A standard candle is one which emits a specified number of lumens of light, where the *lumen* is a unit of light quantity. If a source of light is so small that it can be thought of as a mathematical point, then rules of solid geometry enable one to calculate the amounts of light emitted by sources of different candle-power intensities. If a luminous source is not a point but an extended surface, its brightness rather than its intensity is designated. *Brightness* refers to the emission of so many lumens per unit of area.

Illumination. When light falls upon a surface the surface is said to be illuminated. See fig. 14.3. The number of lumens per square foot incident upon a surface is a measure of the *illumination,* and this quantity is usually expressed in foot-candles. The *foot-candle* is the amount of light that falls upon a square foot of surface one foot away from a point source of one

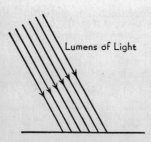

Fig. 14.3. The illumination of a surface is measured by the number of lumens incident upon a unit of area. One lumen per square foot is a foot-candle and is the amount of illumination received on a surface one foot away from a standard candle.

candle-power intensity. Foot-candle meters have become rather common instruments today with the growth of popular interest in photography. It is a foot-candle meter which the photographer holds up to the light to determine the proper exposure time for a given snapshot

Practical Importance of Photometry. These and other photometric matters are of great concern today to the increasingly important profession of illuminating engineering. Industrialists have become aware of the increase in efficiency of workers when proper conditions of illumination are maintained in offices and factories. Instead of the one to three foot-candles of illumination which prevailed in many factories in the past, proper values up to 40 or 50 foot-candles are maintained for general illumination in the modern factory. The illumination of a surface exposed to full summer sunlight may run as high as 10,000 foot-candles, while under full moonlight this figure may be reduced to a very small fraction of a foot-candle.

Law of Illumination. There is a very important law of photometry known as the inverse square law. This law states that the illumination of a surface by a given point source of light varies inversely as the second power of the distance between source and surface. This means, for example, that if the distance between the two is doubled, the illumination is decreased to one-fourth of its original value. Another form of this law states that for equal illuminations due to two point sources of different intensity, the ratio of intensities is proportional to the second

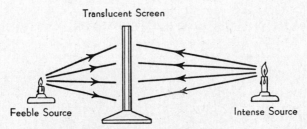

Fig. 14.4. To produce the same illumination on a screen a feeble source must be much closer than an intense source. The illumination produced by a point source is inversely proportional to the second power of the distance from source to screen.

power of the separation, meaning that for a light twice as intense as another, the former must be about one and one-half times as far away as the latter to produce the same illumination at a given place. See fig. 14.4.

Reflection. In the study of geometric optics two phenomena are of the utmost importance. These are *reflection* and *refraction*, each of which is characterized by a basic relationship or natural law. In the case of reflection the so-called law of regular reflection states that the angle made by a reflected beam of light and a perpendicular line drawn to the reflecting surface at the point of reflection is always equal to the angle made by the incident beam and the same perpendicular line. It states, moreover, that

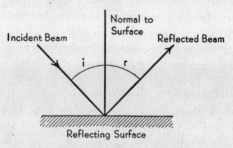

Fig. 14.5. The law of regular reflection states that $r = i$.

the reflected beam, the incident beam, and the perpendicular line just mentioned all lie in the same plane. See fig. 14.5.

Refraction. Refraction is the phenomenon by which a ray of light appears to be bent as it passes from one medium into another of different optical character because of a change in the speed of light from the one medium to the other. The basic law here is *Snell's Law*. This law is usually expressed in mathematical language and refers to the relative bending of light rays in two adjacent media in terms of angles and speed factors; hence the reference to geometric optics again. Although incapable of being precisely expressed without the use of mathematical language, Snell's law states essentially that, with reference to a line drawn perpendicularly to a boundary surface between two media at the point where a ray of light passes from the one to the other, the ray is bent toward this perpendicular if the second medium has a greater refractive index than the first, and away from it if the second medium has a smaller index. By *refractive index* of a medium is meant the ratio of the velocity of light in vacuum to that in the medium. Thus as a ray of light passes from air into glass or water, substances in which it is slowed up, it is so bent that the angle made with the perpendicular in the glass, or the water, is smaller than the angle made by the incident beam and the same perpendicular in air. See fig. 14.6. This is to say that the *angle of refraction* in glass is smaller than the *angle of incidence* in air. Moreover, the angle of refraction, as thus defined, is

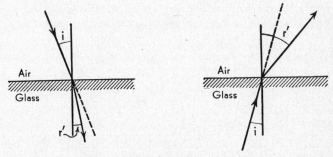

Fig. 14.6. Two cases of refraction, showing the bending of light as it passes from one medium to another.

smaller for glass than for water because the refractive index of glass is greater than that for water.

Although the mathematical aspect of Snell's law need not be emphasized in this presentation, the law itself is expressed mathematically as follows:

$$\mu = \frac{\sin i}{\sin r'}$$

where μ is the relative refractive index of the second medium with respect to the first, i is the angle of incidence, r' is the angle of refraction, and in each case sin stands for the sine of each angle and is a trigonometric function.

The Phenomenon of Total Internal Reflection. An interesting consequence of this law is the phenomenon of *total internal*

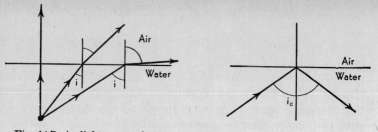

Fig. 14.7. As light passes from water toward air a certain limiting angle of incidence is encountered such that when it is exceeded, all the light is totally internally reflected back into the water. This angle is indicated by i_c in the second diagram.

reflection. When light passes from a medium into another medium of smaller refractive index, there is a limit to the size of the angle of incidence for which light will be refracted. If the angle of incidence exceeds this *critical angle,* as it is called, the light instead of being refracted is all reflected back into the initial medium as if by a mirror, and obeys the law of reflection. See fig. 14.7. It is obvious that as the ray is bent away from the normal under these circumstances it cannot be bent so as to make an angle greater than 90°; hence the limit which is imposed upon the size of the initial angle of incidence. For glass in air this angle is approximately 42°, which is less than the 45° of an isosceles right triangle. Thus light attempting to pass through a right triangular piece of glass parallel to one side will, upon incidence at the oblique side, be reflected out of the glass in a direction parallel to the other side. See fig. 14.8. In other words it will be bent through 90° as by a mirror held at 45° to the incident ray, but without the aid of any silvered material whatever. The bending is said to be the result of total internal reflection.

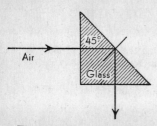

Fig. 14.8. The critical angle for glass-air surfaces is about 42° or less than the 45° incident angle in the diagram. Therefore the light is totally internally reflected as shown.

Images Formed by Plane Mirrors. Considerations of the laws of reflection and refraction lead to numerous conclusions regarding the behavior of rays of light in optical media, accounting

PLANE MIRRORS

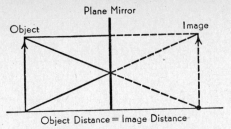

Fig. 14.9. The image formed by a plane mirror is erect and vertical. It is formed just as far behind the mirror as the object is in front of it.

for such things as the formation of images by the use of mirrors, prisms, and lenses. Since all optical instruments depend for their operation upon these considerations, their importance is not to be underestimated. Consider the plane mirror for example. It follows from the law of reflection that a right-side-up image, formed apparently just as far behind the mirror as the object is in front of it, is accounted for. See fig. 14.9. This simple phenomenon is, of course, an everyday occurrence, but nevertheless an important

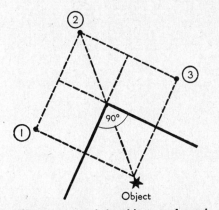

Three images of the object are formed by two plane mirrors making an angle of 90° if the object is placed on the line bisecting the angle between them.

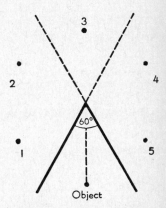

Five images are formed by an object placed as shown between two plane mirrors making an angle of 60° with each other.

Fig. 14.10. Examples of multiple images formed by reflection for plane mirrors.

one from the viewpoint of physics. Other examples may easily be demonstrated. See fig. 14.10.

Reflection from Spherical Surfaces. As rays of light are reflected from curved mirrors the results are even more interesting. A spherical *convex* mirror always makes objects appear reduced in size, but right side up; whereas a spherical *concave* mirror can sometimes produce inverted images standing right out in space in front of the mirror and at other times produce right-side-up images located apparently behind the mirror. See fig. 14.11. Such inverted images are known technically as *real images* because they really exist where the rays of light are focused. They can be localized upon a screen. Many optical illusions are produced by real images. The right-side-up images are called *virtual* images. These, however, can never be "captured" on a screen.

It is obvious that the formation of images by curved mirrors, cylindrical as well as spherical, can be of considerable practical importance in optical instruments.

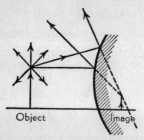
Image by a convex spherical mirror.

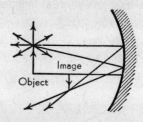
Image (one case) by a concave spherical mirror.

Fig. 14.11. Images by spherical mirrors.

Focal Points. The largest astronomical telescopes in the world, such as the one on Mt. Palomar in California (200 inches across), operate on the principle of the formation of real images of heavenly bodies by large concave mirrors. One of the characteristics of these instruments is the fact that all rays of light from distant objects are brought to focus at a single point known as the *focal point*. See fig. 14.12. The distance from this point to the mirror is called the focal distance. It can be shown that for spherical mirrors the focal distance is just half the radius of

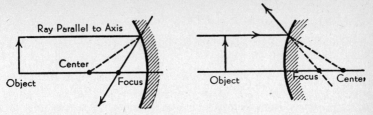

Fig. 14.12. Illustrating the focal point of the concave spherical mirror and the virtual focal point of the convex spherical mirror. The focal distance is one-half the radius of curvature.

curvature. Whereas the convex spherical mirror always produces virtual images reduced in size with respect to the object, and the focal point is virtual, the type and size of the image produced by the concave spherical mirror depends upon the distance between object and mirror as compared with the focal length and radius of curvature of the mirror. The concave mirror is an example of a *converging system* because it actually converges light. The convex mirror, on the other hand, always makes light appear to diverge and hence is called a *diverging mirror*.

Simple Lenses. Another example of a converging system is the double convex lens which, by refraction, causes light to be bent more at the edges than at the center and so to be converged.

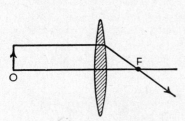 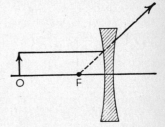

Fig. 14.13. The converging lens has a real focal point toward which rays originally parallel to the axis pass after refraction.

Fig. 14.14. The diverging lens has a virtual focal point from which rays originally parallel to the axis appear to come after refraction.

The double concave lens, on the other hand, being thicker at the edge than at the center, produces the opposite effect and so is an example of a diverging system. See figs. 14.13 and 14.14.

The convergence of a system can be shown by passing a beam of sunlight, restricted by an aperture, through the system if it is a lens, or by reflecting it from it if a mirror, in a smoke-filled room. It will be noted that originally parallel rays (parallel because the sun is about 93,000,000 miles away and a narrow beam does not have much opportunity to spread in the short space of an ordinary room) either converge to a focus or diverge from an apparent focus as indicated. See again figs. 14.13 and 14 14. It

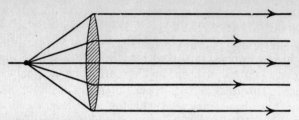

Fig. 14.15. Point source at focus of converging lens. The beam which is transmitted by the lens is a parallel beam.

will also be noted that the rays from a small pointlike source situated on the principal axis at the focal distance away from a converging lens will be parallel upon emergence from the lens. This provides a means of producing parallel light in the laboratory without resort to sunshine. See fig. 14.15.

With a converging system, i.e., either a double convex lens or a concave mirror, it is found that images of objects lying farther away than the focal length are always inverted and real. For objects located within the focal distance, however, images are virtual, erect, and usually enlarged. The size of the image is to the size of the object as the image distance is to the object distance.

Ray Diagrams. From a consideration of the laws of reflection and refraction and a knowledge of the characteristics of the focal points of a simple lens or spherical mirror it is possible to locate the image of an object by a so-called ray diagram.

Ray Diagram for Concave Mirror. Consider an object in the form of an arrow whose base rests on the axis of a concave spherical mirror. See fig. 14.16. Treating the arrow as a luminous source of light, consider first the rays of light emitted from the head end of it. Of all the rays emitted from this point there are

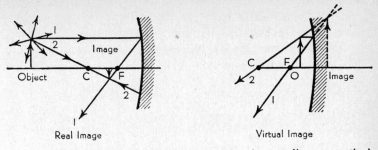

Fig. 14.16. Two examples of images located by the ray diagram method. These are formed by converging mirrors (concave).

at least two whose directions after reflection from the mirror are determinable. The ray which travels parallel to the axis must be reflected through the focal point, and the ray which passes through the center of the curvature must be reflected back upon itself because such a ray is a radial line which by definition is perpendicular to the mirror surface and therefore makes a zero angle of incidence. Except when the base of the arrow is at a distance equal to the focal length away from the mirror, these two rays will intersect in front of the mirror, or their extensions will appear to intersect behind the mirror, and such an intersection will represent an image of the point from which the rays originally emanated. If the intersection takes place in front of the mirror the image is real; if behind the mirror, then the image is virtual. Now it is found that if the object lies outside the focal point the image is *always* real and inverted. If, however, the object lies inside the focal point, i.e., between the mirror and the focal point, the image is virtual and erect. It will be recalled that when we look into a shaving mirror at very close range the observed image is right side up, behind the mirror, and usually magnified. If, however, the mirror is held at quite some distance, the image of the face is wrong side up, or real. This is because the focal distance has been exceeded.

Ray Diagram for Convex Mirror. The same ray-tracing procedure can be followed with the convex mirror, only to discover that in this case the image is always virtual and reduced in size. See fig. 14.17.

An interesting check on both these cases results when a third ray is considered. Obviously the ray from the head end of the

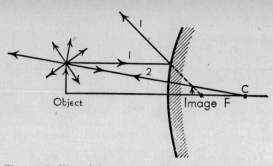

Fig. 14.17. Virtual image by a diverging mirror (convex) located by ray diagram method.

object which passes directly through the focus must be reflected parallel to the axis and it should also pass through the point of intersection of the other two rays.

Of course, it would be necessary to consider such rays as are here called first, second, and third for every single point of the object to reproduce every single point of the image, but this is hardly necessary since the head alone locates the position of the image when the base is on the axis.

Ray Diagrams for Thin Lenses. The foregoing considerations of rays is applicable also to thin lenses except that instead of being reflected the rays are transmitted through the lens, and instead of considering the ray through the center of curvature, there being no such thing for a two-sided lens, the ray passing through the actual center of the lens is used. See fig. 14.18. It can be shown that this ray is undeviated after passing through a thin lens. By such an analysis it is found that diverging lenses

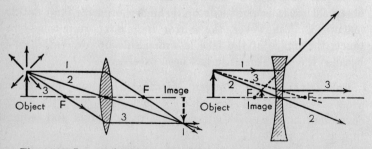

Fig. 14.18. Images found by ray diagram method for converging and diverging lens. These show how a third ray through the focal point emerges parallel to the axis and checks the result given by rays 1 and 2.

always produce virtual and diminished images whereas converging lenses produce real images if the objects lie outside the focal point and virtual images if the objects lie within this distance.

The Magnifying Glass. A common example of the latter situation is that illustrated by the ordinary reading glass or

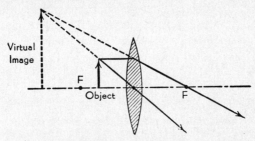

Fig. 14.19. The simple microscope or ordinary magnifying glass represents a special case of image formation by a converging lens when the object is placed between the focal point and the lens. The image is virtual and on the same side of the lens as the object.

simple microscope as it is technically called. In use, the glass is held at a distance from the printed page less than the focal length of the lens and the resulting image is right side up (virtual) and on the same side of the lens as the object itself. See fig. 14.19. If the glass is held too far away from the printed page the image of the print is inverted (real) and is located out in front of the lens, i.e., on the opposite side from the object.

Almost all optical instruments are combinations of mirrors and lenses, and in all cases it is possible to trace rays of light through the instruments to understand how they function. Such instru-

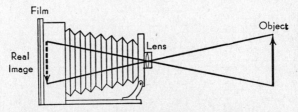

Fig. 14.20. The action of the photographic camera is illustrated. A real image of the object is formed on the film placed at the rear of the camera.

ments include telescopes, microscopes, opera glasses, cameras, and the all-important human eye. See fig. 14.20. For an optical system consisting of more than one mirror or lens it is proper, in ray diagram considerations, to first locate the image formed by the first unit of the system and then treat it as the object for the second unit, and so on in succession.

The Eye. Possibly the most important optical instrument is the human eye, although the mistake must not be made of confusing the phenomenon of "seeing" with that of the image formation by the optical system of the eye. Seeing involves the interpretation by the brain of the image that is formed. Here we deal only with the eye as an optical instrument, however, a sort of camera in which an image of an object is formed on a nervous tissue at the back of the eye, called the *retina*, which plays the role of the film in the photographic camera. In the front of the eye a combination of an aperture called the *pupil* and a lens arrangement serve as an optical system to produce images of outside objects. These images falling on the retina, which is really the inner lining of the rear part of the eyeball, stimulate the nerve cells contained in it to send appropriate messages to the brain. See fig. 14.21. The whole eye is filled with water-like fluids which also take part in the refraction phenomena. The aperture, or pupil, is controlled in size automatically by the action of the iris diaphragm. The smaller the opening, the sharper the image that results; but too small an opening allows insufficient light to enter to stimulate the retina, and so the adjustment of the size of this opening, although performed automatically, is indeed very critical.

One of the important features of the eye is the focusing arrangement. Muscles are so attached to the lens that it can be flattened or bulged, with the focal length correspondingly altered.

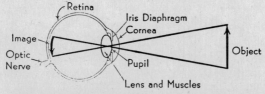

Fig. 14.21. The optical system of the human eye serves to form real images on the retina, the nerve tissue lining the back portion of the eyeball.

It is by this arrangement that the eye is said to be *accommodated* for objects at different distances. Otherwise it would not be possible to focus the eyes first upon a distant scene and immediately thereafter upon a newspaper held at ordinary reading distance. It is this power of accommodation which is gradually lost with old age, and this loss necessitates the use of so-called "double-vision" glasses. Indeed the entire question of spectacles is an application of physics in everyday life which will be considered presently.

Although the whole retina is very sensitive to light, there is one spot in particular that is extremely sensitive. In fact it is so sensitive that in all close work such as reading, for example, the eyeballs are continually moved so as to focus the work on these spots of the two eyes. Indeed, these spots are so small that it is necessary to actually move the eyeballs to shift the attention from the one to the other of the two dots which make up the semicolon mark of punctuation. Thus it is that reading makes it necessary for the eyes to be moved jerkily along the lines of the printed page, focusing upon one spot after another in rapid succession. Is it any wonder that these eyes of ours sometimes get tired and require the assistance of spectacles, especially in middle life or later, for many people?

Eye Defects and Spectacles—Nearsightedness. It has already been suggested that the adaptation of spectacles to assist human vision is a practical application of physics. Some people are born with eyeballs too long for distinct vision. Images formed by the lens system fall short of the retina, and no matter how

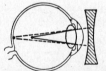 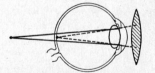

In myopia (nearsightedness) the image is formed before the retina. With a diverging lens in front of the eye the image is formed on the retina. Thus faulty vision is corrected.

In hypermetropia (farsightedness) the image is formed behind the retina. With a converging lens in front of this eye the image is formed on the retina.

Fig. 14.22. Eye defects.

hard the eye attempts to accommodate, distinct vision is **not** achieved except for very close objects. This condition is popularly called nearsightedness. Technically it is referred to as *myopia*. The obvious remedy is the use of a diverging lens in front of the eye to increase the focal length of the whole system. This is, then, why nearsighted persons wear glasses which are thicker at the edges than in the centers. These are diverging glasses. See fig. 14.22.

Farsightedness. The opposite condition is of course farsightedness, or *hypermetropia*, as it is called. It represents a condition in which the eyeball is too short and, as a result, images are always formed behind the retina. This condition is relieved by the use of supplementary converging lenses to shorten the overall focal length of the system. As people lose the power of accommodation with increasing age they generally become more and more farsighted. The near point, so to speak, recedes with advancing age. Thus it is that many a nearsighted person finds his vision improved as time goes on.

Astigmatism. Many eyes suffer from a defect known as astigmatism. It is probably safe to say that more spectacles are worn for this reason than for any other. Astigmatism is an aberration involving the shape of the front surface or the *cornea* of the eye. This defect shows up when a person attempts to see with equal distinctness all the lines of a series drawn radially from a point in a plane. The astigmatic eye sees the lines in certain directions more distinctly than in others. Headaches constitute a common symptom of this difficulty as a result of strain on the eye muscles when they attempt to compensate for the defect. To correct a condition of astigmatism cylindrical spectacles are used with their cylindrical axes properly adjusted for the direction of the astigmatism. Of course several defects can be compensated for at the same time by a pair of spectacles so that spectacle lenses are often combinations of spherical and cylindrical surfaces, representing something of an achievement on the part of the optician. Indeed, it is no exaggeration to repeat that applications of physics in the aid of human vision are of the utmost importance.

Summary. In this chapter we have seen that geometric optics plays a very important role in the world of human experience.

From the ordinary spectacle lens to the largest astronomical telescope, light rays form images of luminous objects. Reflection and refraction, mirrors and lenses, illumination and darkness—all are within the realm of optics, all a part of the study of physics. Let us consider next the action of a prism on a beam of light, the wonders of the rainbow, and other optical phenomena related to physical optics as contrasted with geometric optics.

QUESTIONS

1. What is meant by photometry?
2. Does the intensity of illumination at a place depend upon the distance it is away from the source?
3. Define the foot-candle.
4. Why can it be assumed that light passes through the center of a thin lens almost undeviated in direction?
5. Indicate the path of two rays from a point on an object which may be drawn to locate the image formed by (a) a spherical mirror; (b) a lens.
6. Show by a diagram that a person can see his entire image in a plane mirror one-half as tall as he is.
7. How can farsightedness and nearsightedness be corrected with spectacles?
8. What type of lens must be used to produce an enlarged image on a screen?
9. Show by a diagram how light passes (a) through a lens which is thicker at the center than at the edges; (b) through a lens whose edges are thicker than the center.
10. Can a double convex lens be used to start a fire? A convex mirror? A concave mirror?
11. What type of image is formed on the film of a camera by the lens?

CHAPTER XV

OPTICAL CONSIDERATIONS (*Continued*)

PHYSICAL OPTICS, DISPERSION, SPECTROSCOPY, INTERFERENCE, DIFFRACTION, POLARIZATION

In the preceding chapter, questions dealing with the directional aspects of light rays were considered. It was observed that rays were reflected, bent, made to intersect one another to form images, and affected by the presence of matter. The geometry of each situation was of more immediate importance than the nature of light itself. In this chapter we are to be more concerned with the latter question. We shall first consider another phenomenon in which direction also plays an important part.

The Visible Spectrum. When a narrow beam of sunlight is passed through a triangularly shaped glass prism as indicated in fig. 15.1 a very interesting and perhaps strange result is produced by refraction. Not only is the ray bent, as the study of the preceding chapter would require, but it is spread out into a whole spectrum of visible colors ranging from red to violet. This indicates that there is something about a beam of light which makes possible its resolution into component parts. The phenomenon is referred to as *dispersion*. If the source of light is an illuminated narrow slit and an image of it is formed on a screen by a lens held near the prism, there will be reproduced on the screen successive overlapping images of the slit in every single color present in the source. Thus the spectrum formed by a prism is characterized by the light source used, and many different types of spectra have been classified according to the nature of the source.

Spectroscopy. All incandescent light sources are found to emit continuous spectra, i.e., continuous bands of light containing all the colors of the rainbow. Indeed, the rainbow itself is a spectroscopic phenomenon produced by sunlight passing through droplets of water in the air when the sun makes a certain angle with the horizon after a shower. The solar spectrum, however, is not strictly a continuous spectrum, as will be pointed out presently. On the other hand, electric-light filaments and glowing furnaces do give spectra which are continuous.

Hot vapors, like those in the modern fluorescent lamps, give rise to spectra in which only selected colors are present. Again,

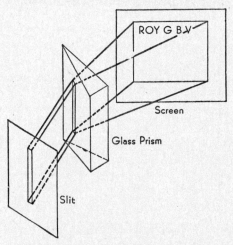

Fig. 15.1. A beam of sunlight is spread out into a colored spectrum by a triangularly shaped glass prism.

if the source is a slit illuminated by such a light, the spectrum will be a set of discrete lines separated by dark gaps. Such a spectrum is ordinarily referred to as a *bright-line* spectrum. No two chemical elements yield the same bright-line spectrum, whence identification of chemical substances is possible by spectroscopic methods. This is in accordance with our knowledge of the structure of the atom (see chapter XIII). For a given atom such as the atom of hydrogen, and considering the Bohr model of its structure, as energy is added, the revolving electron jumps to an outer orbit. As it falls back from some outer orbit to some inner orbit, energy is released in the form of radiation of a given frequency according to the quantum principle, $E = h\nu$ (where E stands for energy, h is Planck's constant, and ν is the characteristic frequency). If this energy is in the visible frequency range, then it is made evident by a bright spectral line. Indeed all the lines in the hydrogen spectrum are thus accounted for. Similarly the characteristic bright lines appearing in the hot vapor spectra of all the chemical elements are also explained, whence it is clear that no two chemical substances have exactly the same spectral patterns. Moreover it is because of extensive studies in the spectroscopy of the elements that today so much is known about the structure of matter.

The third type of spectrum in this classification of spectra is the *dark-line* or *absorption* spectrum. This type is produced where an incandescent source is placed in such a way with respect to a hot vapor that the light from the former is intercepted by the latter before it passes through the spectroscope. Under these conditions the colors which would ordinarily constitute the spectrum from the hot vapor are absorbed from the continuous spectrum that would have been produced by the incandescent source alone. This results, in the case of the slit-type spectroscope, in an almost continuous spectrum, but one crossed by dark lines.

Fraunhofer Lines. The solar spectrum is a dark-line spectrum and indicates that the outer portions of the sun contain elements whose characteristic wave lengths have been absorbed from the otherwise continuous spectrum of the inner and hotter portions of the sun. The dark lines in the solar spectrum, having first been observed by the German physicist Fraunhofer, have been called *Fraunhofer lines*. The more conspicuous of these are labeled with alphabetical letters. See fig. 15.2. Thus the Fraunhofer D line refers to the sodium doublet, or the pair of very close lines in the yellow-orange region of the spectrum which occurs at precisely

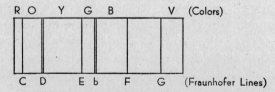

Fig. 15.2. The appearance of the solar spectrum crossed by the Fraunhofer lines. The colors in this spectrum are unequally spaced when formed by a prism. The D line is due to sodium, the C line to hydrogen, etc.

the same position as that occupied by a similar pair of yellow-orange lines in the spectrum of a sodium vapor lamp. The so-called optical dispersion of a spectroscope is often measured by the degree to which the instrument can separate the lines in this doublet.

Discovery of Helium. It was because certain Fraunhofer lines could not at first be identified that the existence of a new element on earth was postulated. It was even named helium, the sun element, long before its presence on earth was later discovered after a conscious search was undertaken.

Thus the whole question of the nature and structure of light is raised by the study of spectroscopy. It is evident that so-called white light, or simply sunshine, can be broken down into constituent colors by the prism of the spectroscope. Moreover, it is possible to recombine colors so as to reproduce white light. These are simple experimental procedures which can be performed quite readily in the classroom today just as Newton demonstrated them some three hundred years ago.

Color Mixtures. This matter of the superposition of colors or the mixtures of colors is an interesting study in itself. It provides a physical basis for art. It is found that only three colors, the so-called primary colors, are necessary to produce a fairly accurate reproduction of white light, and that combinations of these three colors will reproduce practically any color. Color printing utilizes three primary colors in this manner. In a less accurate way, two colors sometimes produce a fairly good reproduction of white light. Such pairs of colors are called complementary colors. Blue and yellow, red and green, are examples of such complementary pairs.

Color Addition vs. Color Subtraction. An important distinction must be made between the addition of two colors and the addition of two pigments, which is really the subtraction of two colors. Yellow and blue light when added together produce white light, as just noted, but yellow paint applied to a blue surface produces a green surface. Since the color of yellow paint results from the fact that all the colors but yellow, probably bordered by some green, have been absorbed, and since the color of blue paint results from the fact that all the colors except blue, probably bordered by some green, are absorbed, the only color reflected by the combination is that which is reflected in common by each, namely green. Thus the addition of yellow pigment to blue pigment is the subtraction of all colors except green.

This distinction between addition and subtraction can be shown in another way. If two spotlights, one with a yellow filter before it and the other with a blue filter, are caused to shine their light upon the same screen, the resulting spot of light on the screen will be practically white. If, however, a single spotlight is used with both yellow and blue filters, the resulting spot on the screen will be green. This is because the yellow filter prevents all but yellow light from passing, although the cut-off wave length will probably not be very sharp and some green will be passed. Similarly the blue filter prevents all but blue light from passing, but probably some green will get by because of incomplete filtering. The only color, then, that will probably get through both filters is green. Of course, a complete filtering job by the filters would produce black, or the complete absence of all color. It might be noted parenthetically that no object has any color in the dark. Color is the result of selective absorption of light. Similarly, a red apple appears black under strictly green light because there is no red light to be reflected if the light is truly green.

Interference. The superposition of light not only involves the matter of additive colors but also suggests another phenomenon of considerable interest. Under very special circumstances light apparently coming from two different sources can be made to produce darkness; that is to say, two beams of light can be made to nullify each other. Thomas Young (1773–1829) observed that when a candle is placed behind a double slit, i.e., two parallel slits rather close together, the light coming from the two slits forms a peculiar pattern on a screen held in front of the slits. This pattern consists of a set of alternate bright and dark bands oriented parallel to the slits. When white light such as a candle source is used, the bright bands are colored, but when light of a single color is used, e.g., light from a yellow sodium source, the bright bands, or fringes as they are called, are all of one color, separated of course by dark bands. (See the diagramatic representation of Young's experiment, fig. 15.3.)

The explanation of this phenomenon involves the wave theory of light. Whereas water waves are known to be able to interfere

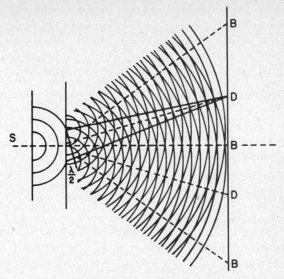

Fig. 15.3. Young's interference experiment.

with one another when they are superimposed if one is a half wave length out of phase with the other, so it follows that light waves can interfere destructively under proper circumstances. The dark bands, D, D, etc. occur at those places where the distance to one slit exceeds the distance to the other slit by an integral number of half wave lengths. Of course the distances between the fringes are very small since the wave length of light is very small, being of the order of several hundred thousandths of a centimeter, and the separation of the slits must be reasonably small for the same reason. The whole phenomenon is called interference of light. The dark bands represent destructive interference and the bright bands represent constructive interference.

Diffraction. Although the prism spectroscope provides a means of breaking light down into its component parts, and in particular of separating white light into colors by refraction, there is still another way of producing the same effect. This is by *diffraction*, which is a phenomenon of selective bending of light rays according to wave length as light passes around the edges of obstructions, the longer waves being bent more than the shorter ones. A grating produced by ruling several thousand lines per inch on a transparent glass plate by a diamond point constitutes a device by means of which white light can be thus resolved into its colors. As the light passes through this grating some of it is bent in direction, with red light bent more than blue light. Consequently, spectra are produced by a fine-ruled grating. See fig. 15.4.

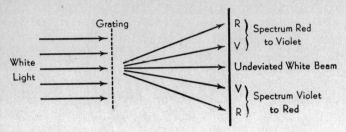

Fig. 15.4. A grating formed by ruling several thousand lines per inch on a glass plate will produce spectra by diffraction as indicated.

This is in reality an interference phenomenon and requires for its interpretation the assumption that light is a wave phenomenon, for corpuscular light would not be so affected by a grating. As has been noted, the bending of light is selective according to wave length.

Grating Spectroscope. This so-called grating type of spectroscope makes possible the very simple matter of measuring the wave lengths of light. The bending of waves around edges is very familiar to all who have watched water waves play around docks and piers. What is not so familiar is the manner in which the bending depends upon the length of the waves concerned. Very short light waves (.00004 cm. to .00007 cm.) are ordinarily bent so little that sharp shadows are encountered and the propagation of light is said to be rectilinear; i.e., light travels in straight lines. Under the somewhat extraordinary conditions of the grating, however, the selective bending is exaggerated and a spectrum is formed. To be sure, a ruled grating is so expensive that this phenomenon of diffraction might seem incapable of demonstration except under very favorable circumstances. Such, however, is not the case because inexpensive replicas of costly ruled gratings can be made very easily out of collodion that work practically as well as the originals themselves.

Colors of Thin Films Due to Interference. Another interesting phenomenon in this connection is the coloring produced by very thin films such as thin films of oil on water, or those films which constitute the so-called mother-of-pearl in oyster shells. This also is an interference effect and one which can be produced only when the films are exceedingly thin. The light reflected from one surface interferes with the light reflected from the other surface; re-enforcement or nullification results depending upon the number of half wave lengths included between the two surfaces. Indeed, the thickness of the film can sometimes be measured by the color effects produced. This is a way of utilizing light waves for measurement purposes. Such interferometric methods are used in precision measurements to such extent and refinement today that it is possible to measure to a millionth of an inch with certainty. Thus is made possible the interchange of working parts in the modern automobile, which have to be fitted with as little clearance as a few thousandths of an inch.

Invisible Glass. A very novel and interesting application of the laws of interferometry has been developed to make glass invisible. This is accomplished by coating the glass with a film of some convenient substance. The thickness of this film is a very critical matter; it is just so thick that light reflected from one side of it will interfere destructively with light reflected from the other side—(not over a quarter of a wave length of light at the most), thereby eliminating the beam that would normally be reflected from the front surface of the glass. This, of course, makes the glass invisible since the line of demarcation between it and the surroundings is eliminated. It is to be noted that the actual surface of a front-surfaced mirror is really invisible since the reflection is total rather than selective. Camera lenses coated so as to make their surfaces invisible are found to be very effective in producing glareless photographs. Similarly, invisible-glass windows in front of instruments such as clocks and electrical meters cut down the glare to observers, who can thereby see the faces of the instruments more clearly.

Polarized Light. As a final topic to be considered in this descriptive survey of optics, and of physics generally, we turn to polarized light, a subject which has gained considerable popularity in recent years through the advent of a polarizing material used in certain sunglasses and in three-dimensional movies. The concepts of interference and diffraction just considered have definitely established the wave aspects of light, and yet it should be recalled from the quantum theory viewpoint that the corpuscular aspects are not to be ignored. Nevertheless it must be obvious that, for the most part, the wavelike features of light seem to be invoked more often to explain common phenomena than the corpuscular features. With this in mind the next question we encounter is whether light waves are longitudinal like sound waves or transverse like elastic waves. This question is best answered by the study of the polarization of light.

When a beam of light is passed through a crystal of the substance tourmaline, or through a polarizing disk, it finds itself in a very special condition. It can only be passed through a second crystal of tourmaline or polarizing disk if the latter is oriented exactly as the first one. If the second polarizing disk, for example, is rotated through an angle of 90° with respect to the first one, no light is passed through it. The phenomenon is called *polarization*. The light is said to be polarized by the first disk, which is called the *polarizer*, and in this condition its angle of polarization can be determined by the second disk, which is called the *analyzer*.

Nature of Polarized Light. To help understand the matter better it is useful to consider an analogy. If a horizontal rope fixed at one end is waved up and down by the other end, a transverse wave vibrating in a vertical plane is established. Furthermore, the situation would not be altered if the rope were to pass through a vertical slot or gate. If, however, such a slot were turned so as to

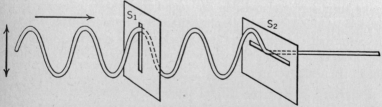

Fig. 15.5. Polarized light. Vertical vibrations of the rope will pass through a vertical slot S_1 but will be stopped by a horizontal slot S_2. The wave pictured is polarized; i.e., the vibrations are in one plane only. Light can be so polarized by certain substances such as tourmaline.

be horizontal, it would effectively stop the wave motion beyond the point on the rope where it was placed. See fig. 15.5. Moreover, if two slots were used, one at some distance beyond the other, then a random sidewise motion of the free end of the rope would be restricted by the first slot to a motion parallel to itself and, unless the second slot were oriented in the same direction as the first, no wave motion would pass through the second slot. There is something about the molecular or atomic arrangement in the substance tourmaline which restricts waves of light to a particular plane of vibration just as the slots in the above examples restrict the motion of the rope. When the mode of vibration of light waves is so restricted, the light is said to be plane-polarized. Thus it is seen further that the polarizability of light requires that light be a transverse wave motion if it is a wave motion at all, because longitudinal waves cannot be polarized.

Polarization by Reflection. In addition to the phenomenon of polarization by transmission as displayed by tourmaline, light is polarized to a large extent just by reflection alone. Indeed, all reflected light is polarized somewhat, even the glare from a cement highway on a bright day. Thus polarizing sunglasses effectively reduce glare because they eliminate the polarized component of the reflected light. Sky light is also polarized because all the fine particles of water vapor, dust, etc. in the air reflect light. Such reflection, or scattering, of short wave-length light is much more pronounced than that of long wave-length light and so the light reflected from the sky is blue. Although the midday sky is blue, the sunset is reddish. This is because the light from the sunset is transmitted through the earth's atmosphere and appears red because the blue has been removed by scattering. The sky at noon is seen by reflected light, whereas the sunset is seen by transmitted light. Both the scattered and the transmitted components are polarized.

Photoelasticity. Polarized light has been found to be of considerable value to the machine designer. This is because many substances which are transparent to ordinary light have quite a different appearance in polarized light. Furthermore, many substances, ordinarily transparent, under polarized light become opaque when strained. Glass and celluloid, for example, display regular-colored interference patterns when strained and viewed under polarized light. Celluloid models of gears and other parts of machinery will show the regions of stress when viewed in operation under polarized light, and many industrial concerns carry out such photoelastic tests on new designs before going into the production of machine parts and manufactured products. In the aviation industry, in particular, where close tolerances in weight are very important, such photoelastic analyses are utilized to prevent a sacrifice in strength along with weight. Indeed, the subject of polarized light is a branch of physics which illustrates quite aptly the manner in which physics has been consciously put to work by practical men who no longer are content with just engineering experience but who are quick to take advantage of every new development in the scientific laboratory.

In Conclusion. No wonder this is a scientific age, and no wonder laymen are interested in the intricacies of physics even if they do not, and in some cases cannot, delve deeply into the mathematics of the situation. It is for students, and people generally, who are thus sincerely interested in a sober treatment of physics on the descriptive level that this text has been written. It will not, and cannot, provide the necessary comprehension for the person who wishes to utilize the material in a professional way, but it is the hope of the writer that the reader will have had his interest aroused, and will have developed a sense of appreciation of what physics is and what physicists do. Hopefully also, he will have acquired a technical vocabulary sufficiently complete to enable him to read with pleasure and understanding the ever-increasing popular-science literature.

QUESTIONS

1. Sketch the passage of a ray of light through a triangular prism.
2. What kind of spectrum is obtained from moonlight? from starlight?
3. Describe the appearance of a red card as it is moved through a continuous spectrum.
4. Why does a mixture of all pigments produce black whereas a superposition of all colors produces white?
5. Is red light diffracted more or less than blue light? How does this compare with refraction?
6. What is meant by diffraction?

7. What is meant by destructive and constructive interference of light waves?
8. How does the phenomenon of polarization indicate light to be a transverse wave phenomenon?
9. Why does "millionth of an inch" precision suggest optical methods of measurement to the physicist?

REVIEW QUESTIONS
(See page 218 for answers.)
Chapters XII, XIII, XIV, and XV

1. The twitching of a frog's legs led to the discovery of the electric current by: 1. Volta; 2. Ampere; 3. Coulomb; 4. Galvani; 5. Oersted .... ( )
2. The friction in a water pipe is analagous to: 1. electric current; 2. potential difference; 3. electrostatic charge; 4. electric resistance; 5. electromotive force .... ( )
3. Electric resistance in an electric circuit plays a role similar to which of the following in a water system?: 1. pump; 2. pipe; 3. water wheel; 4. friction; 5. pressure gauge .... ( )
4. 1 kilowatt hour is represented by: 1. a 10 watt bulb burning for 1 hour; 2. a 100 watt bulb for 1 hour; 3. a 10 watt bulb for 10 hours; 4. a 100 watt bulb for 10 hours; 5. a 100 watt bulb for 100 hours ...... ( )
5. Electric current is determined by: 1. the total charge that flows between two points; 2. the distance that the charge flows; 3. the time it takes the charge to flow; 4. the charge divided by the time; 5. none of these .... ( )
6. The electric motor depends for its operation upon. 1. heating effect; 2. photoelectric effect; 3. thermionic emission; 4. side thrust of a current-carrying wire in a magnetic field; 5. chemical effect .... ( )
7. Electric power is measured in: 1. amperes; 2. volts; 3. watts; 4. joules; 5. ohms .... ( )
8. The negative electron was discovered by: 1. Franklin; 2. Newton; 3. Coulomb; 4. Thomson; 5. Galvani .. ( )
9. The nucleus of an atom is thought to be composed of: 1. neutrons and electrons; 2. protons and neutrons; 3. alpha particles and electrons; 4. electrons and protons; 5. protons and alpha particles .... ( )
10. The emission of electrons when light falls on a metal surface is called: 1. thermionic emission; 2. Edison effect; 3. X rays; 4. photoelectric effect; 5. fission .... ( )

11. Which of the following terms is most descriptive of the phenomenon utilized in the operation of the hydrogen bomb?: 1. fission; 2. fusion; 3. electrolysis; 4. interference; 5. induction .... ( )
12. X rays are: 1. very short wave length electromagnetic radiations; 2. very long wave length electromagnetic radiations; 3. green radiations; 4. high speed electrons; 5. low speed electrons ...... ( )
13. Photoelectric effect is the emission of: 1. electrons from an illuminated surface; 2. electrons in a gas discharge tube; 3. electrons from a hot body; 4. electrons from a radioactive substance; 5. alpha particles from a radioactive substance .... ( )
14. Which of the following are the same as the nuclei of helium?: 1. X rays; 2. heat rays; 3. gamma rays; 4. beta rays; 5. alpha rays .... ( )
15. Roentgen is best known for his work with: 1. the discharge tube; 2. X rays; 3. 2-element vacuum tube; 4. 3-element vacuum tube; 5. the measurement of light .... ( )
16. Light is thought to be a wave motion principally because of :1. its interference effect; 2. its enormous velocity; 3. its action on the human eye; 4. Einstein's theories; 5. its propagation in straight lines ... ( )
17. Diffraction is: 1. the same as refraction; 2. the same as fluorescence; 3. governed by the law of reflection; 4. the cause of red sunsets; 5. the bending of light around edges .... ( )
18. The age-old conflict between the corpuscular and the wave theories of light has been settled by present-day physicists: 1. in favor of the wave view; 2. in favor of the corpuscular theory; 3. by considering light to be neither wavelike nor corpuscular; 4. by a corpuscular view to explain diffraction and a wave view to explain the enormous velocity; 5. by considering light to be both wavelike and corpuscular ... ( )
19. The velocity of light was determined very precisely by: 1. Newton; 2. Galileo; 3. Michelson; 4. Einstein; 5. Huygens .... ( )
20. The shortest plane mirror in which a man can see his entire erect image: 1. is just as long as himself; 2. is dependent upon his distance away from the mirror; 3. is just half his height; 4. is necessarily twice his height; 5. is indeterminate .... ( )

21. The foot-candle is a measure of: 1. brightness of a source of light; 2. intensity of a source of light; 3. solid angle; 4. glare; 5. illumination of a surface ... ( )
22. The illumination produced by a source of light on a surface varies: 1. directly as the distance between them; 2. inversely as the distance between them; 3. inversely as the square of the distance between them; 4. directly as the square of the distance between them; 5. independently of the distance between them .... ( )
23. Light is converged by a double convex lens principally by: 1. refraction; 2. diffraction; 3. reflection; 4. selective absorption; 5. interference .... ( )
24. When a beam of light falls obliquely on a glass surface after passing through air it is: 1. bent toward the perpendicular to the surface; 2. bent away from the perpendicular; 3. undeviated in direction; 4. diffracted; 5. necessarily "focused" .... ( )
25. An oar partially immersed in water appears "broken" because of: 1. refraction; 2. diffraction; 3. polarization; 4. interference; 5. absorption ...... ( )
26. The spectrum of the sun: 1. is an almost continuous band of colored light covering the whole visible range; 2. is a bright-line spectrum; 3. is produced only by means of a prism; 4. may be produced simply by a concave mirror; 5. is exactly like the spectrum of a mercury arc light .... ( )
27. Real images differ from virtual images in that: 1. real images are always inverted; 2. real images are always erect; 3. real images are sometimes inverted but sometimes erect; 4. virtual images can be impressed on a screen; 5. only converging lenses can produce virtual images .... ( )
28. The wave length of yellow light is approximately: 1. .06 cm.; 2. .0006 cm.; 3. .00006 cm.; 4. .006 cm.; 5. .0000006 cm. .... ( )
29. Yellow light and blue light when added together produce: 1. green; 2. yellow; 3. blue; 4. white; 5. black .... ( )
30. A double convex glass lens is: 1. a diverging system; 2. capable of producing virtual images; 3. unable to produce real images; 4. always free from chromatic aberration; 5. always free from spherical aberration .... ( )

31. The human eye is: 1. essentially a spectroscope capable of analyzing the component colors in a source; 2. essentially a radio receiver permanently "tuned" to a narrow band of electromagnetic frequencies; 3. in no respect whatever a radio receiver; 4. absolutely unaffected by infrared light; 5. essentially a compound microscope .... ( )
32. Myopia (nearsightedness) is corrected by the use of: 1. concave lenses; 2. convex lenses; 3. concave mirrors; 4. convex mirrors; 5. astigmatic lenses .... ( )
33. The largest telescopes in the world are mirrors because: 1. inverted images are desired; 2. they are cheaper and more practical than lenses of the same size; 3. they will gather more light than lenses of the same size; 4. they can be made with longer focal lengths; 5. greater magnification is possible than with lenses of the same size .... ( )
34. Polarized light: 1. was discovered by the use of polaroid; 2. indicates the corpuscular aspect of light; 3. may be used to detect strain in transparent substances; 4. is blue; 5. is of no commercial value as yet .... ( )

Answers to Review Questions pp. 54–58

1. (1)	7. (1)	13. (1)	19. (4)	25. (1)	31. (1)
2. (1)	8. (5)	14. (3)	20. (5)	26. (3)	32. (2)
3. (4)	9. (2)	15. (3)	21. (3)	27. (3)	33. (2)
4. (1)	10. (3)	16. (3)	22. (1)	28. (1)	34. (3)
5. (2)	11. (4)	17. (4)	23. (2)	29. (4)	35. (1)
6. (5)	12. (1)	18. (3)	24. (4)	30. (3)	36. (1)
					37. (1)

Answers to Review Questions pp. 98–99

1. (3)	5. (1)	10. (3)	15. (2)	20. (1)	25. (2)
2. (2)	6. (5)	11. (5)	16. (5)	21. (3)	26. (5)
3. (2)	7. (3)	12. (1)	17. (1)	22. (3)	27. (2)
4. (3)	8. (2)	13. (4)	18. (2)	23. (2)	28. (5)
	9. (4)	14. (1)	19. (1)	24. (3)	

Answers to Review Questions pp. 146–149

1. (5)	6. (3)	11. (1)	16. (2)	21. (5)	26. (3)
2. (1)	7. (2)	12. (5)	17. (2)	22. (4)	27. (4)
3. (1)	8. (4)	13. (3)	18. (5)	23. (3)	28. (2)
4. (2)	9. (1)	14. (3)	19. (4)	24. (3)	29. (2)
5. (3)	10. (2)	15. (1)	20. (1)	25. (3)	

Answers to Review Questions pp. 214–217

1. (4)	7. (3)	13. (1)	19. (3)	25. (1)	31. (2)
2. (4)	8. (4)	14. (5)	20. (3)	26. (1)	32. (1)
3. (4)	9. (2)	15. (2)	21. (5)	27. (1)	33. (2)
4. (4)	10. (4)	16. (1)	22. (3)	28. (3)	34. (3)
5. (4)	11. (2)	17. (5)	23. (1)	29. (4)	
6. (4)	12. (1)	18. (5)	24. (1)	30. (2)	

APPENDIX

SUPPLEMENTARY REVIEW QUESTIONS

1. The scientific method: 1. is based on deductive reasoning; 2. is based on inductive reasoning; 3. is illustrated by the writings of Aristotle; 4. has always been followed in physics; 5. disregards facts ... ( )
2. A hypothesis is: 1. the same thing as a scientific theory; 2. not allowed in physics; 3. never successful; 4. a necessary step in inductive reasoning; 5. always successful .... ( )
3. Force is the same as: 1. mass; 2. push; 3. power; 4. momentum; 5. acceleration .... ( )
4. The resultant of two forces of 10 lbs. each acting in opposite directions is: 1. 20 lbs.; 2. 10 lbs.; 3. 14 lbs.; 4. 5 lbs.; 5. zero .... ( )
5. A force is always necessary: 1. to keep a body moving steadily in a fixed direction; 2. to change the speed of a body; 3. to account for the velocity of a body; 4. to keep a body at rest; 5. to determine the position of a body .... ( )
6. Displacement is: 1. the change of position; 2. the same as distance traveled; 3. the time-rate of change of velocity; 4. the same as momentum; 5. synonymous with magnitude .... ( )
7. Velocity may change: 1. in magnitude only; 2. in magnitude and/or direction; 3. in direction only; 4. only by Newton's law; 5. only horizontally .... ( )
8. The acceleration of gravity is: 1. zero; 2. 32 cm. per sec.; 3. 32 ft. per sec.; 4. 32 cm. per sec. per sec.; 5. 32 ft. per sec. per sec. .... ( )
9. Acceleration is: 1. the same as velocity; 2. the same as displacement; 3. the time-rate of change of displacement; 4. the time-rate of change of velocity; 5. always zero .... ( )
10. A body traveling in a circle with constant speed: 1. has constant velocity; 2. is accelerated; 3. is not affected by gravity; 4. has zero displacement; 5. is weightless .... ( )

11. Newton's second law of motion proposes that: 1. force is mass times acceleration; 2. weight is force; 3. a body at rest continues at rest unless a force acts; 4. to all actions there are equal and opposite reactions; 5. all bodies are attracted to the center of the earth .... ( )
12. Work is: 1. mass times acceleration; 2. force times displacement when these two quantities have the same direction; 3. force times displacement only when the two are mutually perpendicular; 4. force times velocity; 5. mass times velocity .... ( )
13. The work done when a 10-lb. force acts horizontally on a 50-lb. body on a frictionless surface is: 1. 10 ft.-lbs.; 2. 10 lbs.; 3. 500 ft.-lbs.; 4. 500 lbs.; 5. zero .... ( )
14. The capacity to do work is called: 1. mechanical advantage; 2. energy; 3. power; 4. efficiency; 5. momentum .... ( )
15. The ratio of the force overcome by a machine to the force applied to it is called: 1. mechanical advantage; 2. energy; 3. power; 4. efficiency; 5. work ...... ( )
16. The maximum displacement of a vibration from equilibrium is called: 1. frequency; 2. speed; 3. amplitude; 4. period; 5. inertia .... ( )
17. Simple harmonic motion: 1. takes place in a straight line; 2. is rotary; 3. is torsional; 4. is always audible; 5. always involves nodes and loops .... ( )
18. Sound is a wave phenomenon of the following type: 1. transverse; 2. circular; 3. longitudinal; 4. electromagnetic; 5. elliptical .... ( )
19. The velocity of a wave is: 1. the product of the frequency and the wave length; 2. the ratio of the frequency divided by the wave length; 3. mass times acceleration; 4. the distance from a crest to the next crest; 5. always 186,000 miles per second ...... ( )
20. In a tube open at each end sound waves: 1. display a node at either end; 2. display an antinode at the center; 3. resonate with a fundamental wave length of four times the tube length; 4. have a fundamental wave length of twice the tube length; 5. can never resonate .... ( )
21. The number of overtones in a sound: 1. determines the pitch; 2. determines the quality; 3. gov-

erns the loudness; 4. accounts for all beat phenomena; 5. is always zero .... ( )
22. When a 256-cycle tuning fork and a 260-cycle tuning fork are sounded together, the number of beats one hears is: 1. zero; 2. one; 3. two; 4. three; 5. four. [CAUTION! Do not confuse the answer with the number attached to the answer.] .... ( )
23. Whenever two waves of the same frequency, speed, and amplitude traveling in opposite directions are superimposed: 1. destructive interference always results; 2. constructive interference always results; 3. refraction is demonstrated; 4. the phase difference is always zero; 5. standing waves are produced .... ( )
24. Atoms are: 1. rigid spheres; 2. composed of molecules; 3. always stationary; 4. composed of electrons, protons, and neutrons; 5. nonexistent .... ( )
25. If the pressure of a gas is increased threefold at constant temperature the volume: 1. must be increased threefold; 2. must be decreased threefold; 3. must be decreased ninefold; 4. remains the same; 5. decreases to zero .... ( )
26. If the temperature of a gas is raised its molecules: 1. are speeded up; 2. are slowed down; 3. undergo no change in speed; 4. are made smaller; 5. are made larger .... ( )
27. The temperature on the Centigrade scale which corresponds to $-40°$ Fahrenheit is: 1. $0°$ C.; 2. $-22.2°$ C.; 3. $-72°$ C.; 4. $10°$ C.; 5. $-40°$ C. .... ( )
28. Boyle's law has to do with: 1. liquids seeking their own level; 2. the diffusion of gases through porous substances; 3. the law of multiple proportions; 4. Magdeburg hemispheres; 5. the pressure-volume relations of a gas .... ( )
29. Heat is: 1. a fluid; 2. energy; 3. power; 4. momentum; 5. the same as temperature .... ( )
30. Pressure may be measured in: 1. pounds; 2. inches; 3. pounds per inch; 4. pounds per square inch; 5. pounds per cubic inch .... ( )
31. When the atmospheric pressure decreases, the temperature at which water boils is: 1. raised; 2. always $100°$ C.; 3. lowered; 4. always $212°$; 5. unaltered .... ( )
32. The reason for insulating a refrigerator is: 1. to pre-

vent outward radiation; 2. to prevent heat from coming in; 3. to prevent the cold from going out; 4. to melt the ice; 5. to prevent the ice from melting .. ( )
33. When an ebonite rod is rubbed with fur: 1. electricity is generated; 2. positive and negative electricity are separated; 3. the fur is said to be negatively charged; 4. the rod is said to be positively charged; 5. nothing happens .... ( )
34. The proton compared with the negative electron has: 1. the same sign of charge; 2. the same size; 3. the opposite sign of charge; 4. twice the charge; 5. 1/1800 as much charge .... ( )
35. Two like electric charges: 1. attract each other; 2. repel each other; 3. neutralize each other; 4. have no effect on each other; 5. must be neutrons .... ( )
36. When glass is rubbed with silk: 1. electricity is generated; 2. the glass is positively charged; 3. the glass is negatively charged; 4. the silk is uncharged; 5. the silk acquires twice as much charge as the glass .... ( )
37. The leaves of a charged gold-leaf electroscope move apart because: 1. like charges attract; 2. they are positively charged; 3. they are negatively charged; 4. unlike charges attract; 5. like charges repel .... ( )
38. On the electron theory it follows that when an ebonite rod is rubbed with fur: 1. negative electrons are taken away from the fur; 2. positive electrons are accumulated on the fur; 3. protons accumulate on the rod; 4. protons are wiped off the rod; 5. the rod becomes a good conductor of electricity .... ( )
39. The function of a condenser is: 1. to produce a spark; 2. to increase potential; 3. to diminish the charge; 4. to increase the intensity of a spark at a given electrical pressure; 5. to condense a given amount of charge into a smaller volume .... ( )
40. In a metal conductor the two ends of which are kept at a constant difference in potential: 1. a current of electricity is established; 2. electrons flow from positive to negative; 3. protons flow from plus to minus; 4. charge piles up at the negative end; 5. magnetic effects are never produced .... ( )
41. Electrolysis is an effect usually classified as: 1. magnetic; 2. heating; 3. chemical; 4. acoustical; 5. static.. ( )

SUPPLEMENTARY REVIEW QUESTIONS

42. In this locality the north end of a compass needle points: 1. east of north; 2. north; 3. west of north; 4. toward the north magnetic pole of the earth; 5. south .... ( )
43. A unit positive charge at rest and a unit north magnetic pole: 1. attract each other; 2. repel each other; 3. exert no effect on each other; 4. produce other magnetic poles; 5. produce other electric charges .... ( )
44. When an electroscope is charged by induction: 1. the sign of the charge induced is the same as that on the charging body; 2. the sign of the induced charge is opposite to that on the charging body; 3. charges are transferred directly from the charging body to the electroscope; 4. the induced charge is always positive since the leaves diverge; 5. the charged electroscope has the same number of electrons before and after .... ( )
45. When a current is passed through a wire which crosses a magnetic field: 1. the wire is attracted toward the north side of the field; 2. the wire is attracted toward the south side of the field; 3. the wire is not moved in any way; 4. the wire always becomes magnetized; 5. the wire is caused to move at right angles to the field and at right angles to its own direction .... ( )
46. Electromagnetic induction was discovered by: 1. Maxwell; 2. Edison; 3. Einstein; 4. Oersted; 5. Faraday .... ( )
47. Difference in electrical potential is measured by: 1. a voltmeter; 2. an ammeter; 3. a wattmeter; 4. a photoelectric cell; 5. a fuse .... ( )
48. The electric toaster depends for its operation upon which of the following effects of a current: 1. the heating effect; 2. the magnetic effect; 3. the chemical effect; 4. the thermoelectric effect; 5. the electrolytic effect .... ( )
49. Which of the following names is associated with cathode rays: 1. Joule; 2. Franklin; 3. Thales; 4. Oersted; 5. J. J. Thomson .... ( )
50. The following are all modern physicists except: 1. Michelson; 2. Compton; 3. Millikan; 4. Maxwell; 5. Heisenberg .... ( )
51. All of the following terms suggest modern physics

except: 1. quantum theory; 2. relativity; 3. positron; 4. proton; 5. resistivity .... ( )
52. Light is thought to be a wave motion principally because of: 1. its enormous velocity; 2. its ability to be polarized; 3. its propagation in straight lines; 4. its action on a photographic plate; 5. its ability to be focused by a lens .... ( )
53. A foot-candle meter is used to measure: 1. illumination; 2. glare; 3. brightness of a source; 4. candle power; 5. distance .... ( )
54. The bending of light around edges is due to: 1. refraction; 2. reflection; 3. polarization; 4. diffraction; 5. absorption .... ( )
55. The illumination on the flat top of a table 2 feet directly beneath a 40 candle-power lamp is: 1. 10 foot-candles; 2. 10 candle-power; 3. 20 foot-candles; 4. zero; 5. 80 foot-candles .... ( )
56. The visible spectrum of the sun is: 1. a continuous spread of color from red to violet; 2. a spread of color crossed by dark lines; 3. a bright-line spectrum; 4. exactly like the visible spectrum of sodium light; 5. completely lacking in green light .... ( )

Answers to Supplementary Review Questions

1. (2)	10. (2)	19. (1)	29. (2)	39. (4)	48. (1)
2. (4)	11. (1)	20. (4)	30. (4)	40. (1)	49. (5)
3. (2)	12. (2)	21. (2)	31. (3)	41. (3)	50. (4)
4. (5)	13. (5)	22. (5)	32. (2)	42. (3)	51. (5)
5. (2)	14. (2)	23. (5)	33. (2)	43. (3)	52. (2)
6. (1)	15. (1)	24. (4)	34. (3)	44. (2)	53. (1)
7. (2)	16. (3)	25. (2)	35. (2)	45. (5)	54. (4)
8. (5)	17. (1)	26. (1)	36. (2)	46. (5)	55. (1)
9. (4)	18. (3)	27. (5)	37. (5)	47. (1)	56. (2)
		28. (5)	38. (1)		

INDEX

References to names of physicists are not included. For such names, see the chapter on "Historical Considerations," pages 12–27.

Absolute humidity, 110
Absolute temperature, 94, 104
Absolute zero, 94, 104, 105
Absorption of heat, 122
Absorption spectrum, 204
Acceleration, 38, 39; of gravity, 39
Acceptor, 182
Accommodation, 199
Adhesion, 94
Advantage, mechanical, 50
Air, liquid, 112
Alpha particle, 173
Altitude, 69
Ammeter, 161
Ampere, 153, 155, 156
Amplitude of vibration, 61; of wave, 74
Analyzer, 210
Angle of declination, 143; of dip, 143
Anion, 155
Anode, 155
Antiparticle, 178
Arabic system of numbers, 17
Archimedes' principle, 67, 68, 117
Area, 4
Armature, 160
Astigmatism, 200
Atmospheric pressure, 67
Atom, 86, 88, 174, 175, 203
Atomic bomb, 180
Atomic energy, 23, 174
Atomic mass, 179
Atomic nucleus, 177
Atomic number, 179
Attraction, capillary, 95
Average velocity, 37
Avogadro's law and number, 91

Babble chamber, 178

Balance, beam, 10, 35; spring, 7
Balloon, 68
Barometer, 67, 68
Band, energy, 182
Basic concepts, 2
Beats, 75, 76
Beta particle, 173
Bimetallic thermostat, 105, 106
Bohr's atom theory, 87, 175
Boiling, 109, 110
Boiling temperature, 109
Bomb, atomic, 180; hydrogen, 181
Bonding, molecular, 181
Bottle, thermos, 122
Boyle's law, 90
Brakes, hydraulic, 70
Bright-line spectrum, 203
Brightness, 186
British Engineering Units, 9
British Thermal Units, 107, 108
Brownian motion, 89
Buoyancy, 66

Calculus, 19
Caliper, micrometer, 11; Vernier, 11
Calorie, 107
Calorimetry, 107, 108
Camera, photographic, 197; pinhole, 186
Candle power, 186
Capacitance, 136
Capacitor, 136
Capacity, electrical, 135, 136
Capillary attraction, 95
Cathode, 155
Cathode glow, 166
Cathode rays, 167
Cation, 155
Cell, photoelectric, 172

Celsius temperature scale, 113
Centigrade scale, 103
Centimeter, 6
Centrifugal force, 42
Centripetal force, 42
C. G. S. units, 9, 10
Chamber, Babble, 178
Charge, electric, 88, 127
Charge on the electron, 169, 170
Charles's law, 91
Chemical effect of current, 154
Chladni figures, 79
Clock, 9
Clockwise moment, 35; torque, 35
Closed organ pipes, 83
Cloud chamber, Wilson, 178
Coefficient of friction, 52, 53; of heat conductivity, 46; of thermal expansion, 104
Cohesion, 94
Color, 120; addition, 205; subtraction, 205
Colors of thin films, 209
Compass, magnetic, 142
Component of force, 30
Compression, 63, 74
Concave mirror, 192
Concept, 5
Condenser, 136
Conduction of electricity, 133; of heat, 116
Conductor, 182
Conservation of energy, 48
Constructive interference, 75, 76
Continuous spectra, 203, 204
Convection of heat, 117–119
Converging optical system, 193, 194
Convex mirror, 192
Cornea, 200
Cosmic rays, 178
Coulomb, 131
Coulomb's law, 130, 142
Counter, Geiger-Mueller, 178; scintillation, 178
Counterclockwise moment, 35; torque, 35
Critical angle of incidence, 190
Crookes dark space, 166, 167

Current electricity, 150; direction of, 154
Cyclotron, 26, 174

Dark-line spectrum, 204
Day, mean solar, 9
Declination, angle of, 143
Deductive reasoning, 2
Defects, eye, 199
Density, 64, 65
Destructive interference, 75, 76
Deuteron, 175
Device, thermonuclear, 181
Dew point, 111
Diamagnetism, 145
Diffraction, 207
Diffusion, 95
Dip, angle of, 143,
Direction of electric current, 154; of force, 28
Dispersion, 202
Displacement, 36, 37
Distortion, 63
Diverging optical system, 193
Donor, 182
Doppler's principle, 80, 81
Dry ice, 112
Duration, 9
Dynamics, 28
Dyne, 8, 41, 130
Dyne-centimeter, 47

Edison effect, 170
Efficiency of machine, 51
Elastic waves, 72
Elasticity, 59
Electric capacity, 135
Electric charge, 88, 128
Electric conduction, 133, 134
Electric current, 150; direction of, 154; effects of, 154
Electric discharges in gases, 165
Electric field, 131
Electric induction, 138
Electric lines of force, 132
Electric motor, 160
Electric potential, 132
Electric power, 156

INDEX

Electric resistance, 153
Electricity, 87, 127, 150; frictional, 127
Electrolysis, 155
Electrolyte, 155
Electromagnet, 158
Electromagnetic induction, 161
Electromagnetic spectrum, 120
Electromotive force, 152
Electron, 24, 88, 134, 167, 169
Electron volt, 174
Electronics, 23, 164, 169, 172
Electrostatics, 127
Elevator, hydraulic, 69
Emission, thermionic, 170
Energy, 47, 101, 176; nuclear bonding, 179
Energy band, 182
Energy level, 176
Engineering, 9
English system of units, 10
Equilibrium of forces, 33, 34
Erg, 47
Ether, 23
Evaporation, 109, 110
Expansion, coefficient of, 104; thermal, 102, 105
Experimentation, 3
Explanation, 3
Eye, 198
Eye defects, 199

Fahrenheit scale, 103
Farad, 136
Faraday dark space, 166
Farsightedness, 200
Ferromagnetism, 145
Field, electrical, 131
Field intensity, 131, 132
Films, colors on thin, 209
Fission, nuclear, 180
Fixed points of thermometer, 103
Fluid, 63
Fluid pressure, 65
Fluorescence, 167
Focal point, 192
Foot, 6
Foot-candle, 186

Foot-pound, 47
Force, 28; electrical, 87; molecular, 94, 105
Franklin pulse glass, 109
Fraunhofer lines, 204
Freely falling bodies, 39
Frequency, 73; of vibration, 61
Friction, 52, 53
Frictional electricity, 127
Frost, 111
Fusion, heat of, 111; nuclear, 180

Galvanometer, 160, 161
Gamma rays, 173
Gas, 63, 90–93, 106; ideal 92, 106
Gas expansion, 104
Gas pressure, 93
Gaseous discharge, 165
Geiger-Mueller counter, 178
Geissler tubes, 168
General gas law, 90, 91, 106
Generator, 161
Geometric optics, 184
Glass, invisible, 209
Glow, cathode, 166
Gold-leaf electroscope, 137
Gram, 8, 130
Grating spectroscope, 208
Gravitational pull, 7, 48
Gravity, 43, 44, 48

Half life, 179
Heat, 100; of fusion, 111; of vaporization, 108, 112
Heat insulation, 116
Heat transfer, 115
Heating effect of current, 155
Heavy water, 175
Helium, 205
Hooke's law, 60
Horsepower, 49
Hothouse, 123
House heating, 112
Humidity, 110, 111
Hydraulic brakes, 70
Hydraulic elevator, 70
Hydrogen atom, 176
Hydrogen bomb, 181

INDEX

Hydrometer, 67
Hydrostatics, 64
Hygrometry, 110
Hypermetropia, 200
Hypothesis, 3

Ideal gas, 92, 106
Illumination, 186, 187,
Image, optical, 190, 191; real, 192; virtual, 192
Inch, 6
Inclined plane, 51
Index of refraction, 189
Induction, electrical, 138; electromagnetic, 161; magnetic, 144
Inductive reasoning, 2, 16
Inertia, 7, 40
Infrared light, 120
Instantaneous velocity, 37
Instruments of measure, 10
Insulation, heat, 116
Insulator, 134, 182
Intensity of electric field, 131, 132; of magnetic field, 143
Interference, 206; of waves, 75, 76
International meter, 6
Invisible glass, 209
Ion, 155
Isotope, 174; of uranium, 180

Joule, 47
Joule's law, 156

Kilogram, 8
Kilowatt, 156
Kilowatt hour, 156
Kinematics, 28
Kinetic energy, 47
Kinetic theory of matter, 90
Kundt's apparatus, 79

Laser, 182
Law of Boyle, 90; of Coulomb, 130; of gravitation (Newton), 43; of Hooke, 60; of illumination, 187; of Joule, 156; of Lenz, 162; of motion (Newton), 39; of Ohm, 153; of reflection, 188; of refraction (Snell), 188

Left-hand rule, 159, 160
Length, of wave, 2, 5, 73
Lens, 173, 196
Lenz's law, 162
Level, energy, 176
Lever, 51
Lever arm, 35
Light, 184; polarized, 209, 210
Light rays, 185
Lines, Fraunhofer, 204
Liquid, 63
Liquid air, 112
Lodestone, 127, 141
Longitudinal wave, 74
Loops, 77
Loudness of sound, 82
Lumen, 186

Machine, 49
Magnetic compass, 142
Magnetic effect of current, 157
Magnetic field intensity, 143
Magnetic induction, 144
Magnetic lines of force, 143, 144
Magnetic permeability, 145
Magnetic pole, 142
Magnetism, 141; terrestrial, 142
Magnetite, 128, 141
Magnifying glass, 197
Mass, 2, 5, 7, 8; atomic, 179
Matter, 86
Mean solar day, 9
Mean solar second, 9
Measurement, 4, 5, 10, 11
Measuring instruments, 10
Mechanical advantage, 50
Mechanical equivalence of heat, 101
Mechanics, quantum, 177
Melde's apparatus, 80
Mercury, 68
Mercury thermometer, 102
Meson, 178
Meteorology, 69, 119
Meter, international, 6
Metric system of units, 5
Michelson-Morley experiment, 24
Micrometer caliper, 11
Microscope, simple, 197

INDEX

Mirrors, 190–193
M.K.S. units, 10
Molecular bonding, 181
Molecular force, 94, 105
Molecules, 86
Moment of force, 35
Momentum, 40, 41
Motion, 7, 28, 36; Brownian, 89, 90; Newton's laws of, 39–40
Motor, electric, 160
Muon, 178
Myopia, 199, 200

Nearsightedness, 199, 200
Negative electric charge, 87, 129
Negative glow, 166
Neutron, 88, 178
Newton, 8
Newton's law of gravitation, 43, 44; laws of motion, 39, 40
Nodes, 77
North magnetic pole, 142
North-seeking pole, 142
Nuclear bonding energy, 179
Nuclear fission, 180
Nuclear fusion, 180
Nuclear physics, 23
Nuclear reactions, 178
Nuclear transformation, 178
Nucleus, atomic, 177
Number, atomic, 179

Observation, 3
Octave, 82
Ohm's law, 153
Open organ pipes, 83
Optical image, 190–197
Optics, 184
Osmosis, 95
Osmotic pressure, 95
Overtones, 76, 82

Parallel light, 194
Paramagnetism, 145
Particle, anti-, 178; strange, 178
Pascal's principle, 69, 70
Pauli exclusion principle, 181
Pendulum, simple, 62

Period of vibration, 61
Period of pendulum, 62
Permalloy, 144
Permeability, magnetic, 145
Phase, 62, 73, 207
Phonon, 181
Photon, 181
Photoelasticity, 211
Photoelectric cell, 172
Photoelectric effect, 23, 171
Photographic camera, 197, 198
Photometry, 186, 187
Physical optics, 184
Pinhole camera, 186
Pion, 178
Pitch of a screw, 52; of sound, 82
Plutonium, 180
Polarization by reflection, 211
Polarized light, 209, 210
Polarizer, 210
Polarizing material, 209, 210
Pole, magnetic, 142
Position, 36
Positive column, 166
Positive electric charge, 87, 129
Positive glow, 166
Positive rays, 174
Positron, 88, 177, 178
Potential, electric, 132
Potential difference, 133, 134
Potential energy, 48
Pound, 8
Power, 49; electric, 156
Pressure, atmospheric, 67; fluid, 65; gas, 93; osmotic, 95; vapor, 109
Pressure cooker, 109
Primary colors, 205
Principle, Pauli exclusion, 181; uncertainty, 177
Prism, 203
Proton, 87, 88, 175, 178
Pull, 7, 28
Pupil, 198
Push, 7, 28

Quality of sound, 82, 83
Quanta, 123, 176
Quantum mechanics, 177

Quantum theory, 123, 124

Radar, 172
Radiant energy, 119
Radiation of heat, 119, 122
Radio, 172
Radio waves, 120
Radioactive decay, 179
Radioactivity, 22, 24, 173
Rainbow, 202
Rarefaction, 74
Ray, alpha, 173; beta, 173; cathode, 167; gamma, 173; light, 185; positive, 174
Ray diagrams, 194, 195
Rays, cosmic, 178
Reactions, nuclear, 178
Real image, 192
Rectilinear propagation, 185
Reflection, 76, 188, 211; total internal, 190
Refraction, 77, 188
Refractive index, 189
Refrigeration, 111
Regelation, 113
Relative humidity, 110
Relativity, 24, 125
Resistance, electric, 153
Resonance, 61, 77
Resultant force, 29, 30
Retina, 198
Rotary motion, 34

Sailing into the wind, 32, 65
Science, 9
Scientific law, 3, 43
Scientific method, 2, 3
Scintillation counter, 178
Screw, 52
Second, mean solar, 9
Selective transmission, 122
Semiconductor, 182
Shear, 63
Side thrust, 159
Simple harmonic motion, 60, 61
Simple microscope, 197
Simple pendulum, 62
Sine wave, 74

Sling hygrometer, 110
Snell's law, 188
Solenoid, 158
Solid, 63
Solid state physics, 181
Sound, 81; characteristics of, 82; velocity of, 82
Sound movies, 172
South magnetic pole, 142
Specific gravity, 65
Specific heat, 107
Spectroscope, grating, 208
Spectroscopy, 121, 202, 203
Spectrum, absorption, 204; bright-line, 203; continuous, 203; dark-line, 204; electromagnetic, 120; visible, 202
Speed, 37
Spring balance, 7
Standing waves, 77, 83
Stat-coulomb, 131
Stat-farrad, 136
Stat-volt, 133
Strain, 60
Strange particle, 178
Stress, 60
Stretch, 59, 60
Striations, 166
Sublimation, 112
Submarine, 67
Supersonics, 81
Surface tension, 94, 95
Systems of units, 9, 10

Television, 23, 172
Temperature, 100; absolute, 104
Tension, surface, 94, 95
Terrestrial magnetism, 142
Theoretical mechanical advantage, 51
Thermal expansion, 102
Thermionic emission, 170
Thermometry, 100–102
Thermonuclear device, 181
Thermos bottle, 122
Thermostat, bimetallic, 105
Thin films, 209
Threshold of hearing, 82

INDEX

Time, 2, 5, 9
Torque, 35, 160
Total internal reflection, 190
Transfer of heat, 115
Transformations, nuclear, 178
Transistor, 182
Translatory motion, 34
Transverse wave, 74
Twist, 63

Ultraviolet light, 120
Uncertainty principle, 177
Unit magnetic pole, 142
Unit positive electric charge, 129, 130
Units, 5, 8, 9
Uranium, 180
Uranium fission, 180
Uranium isotope, 180

Vacuum, 68
Vacuum bottle, 122
Vapor pressure, 109
Vaporization, heat of, 108
Vector, 28, 30
Vector addition, 30
Velocity, 37, 38; of sound, 82; of wave, 73
Vernier caliper, 11
Vibration, 61

Virtual image, 192
Volt, 133
Voltage, 132
Voltmeter, 161

Water, heavy, 175
Watt, 156
Wave, 72; elastic, 72; longitudinal, 74; standing, 77; transverse, 74
Wave characteristics, 73
Wave frequency, 73, 74
Wave interference, 75, 76
Wave length, 73
Wave motion, 73
Wave representation, 74
Wave velocity, 73
Weather, 69, 119
Wedge, 52
Weight, 7, 8
Weight density, 65
Weightlessness, 7
Wilson cloud chamber, 178
Work, 46

X ray, 22, 120, 168

Yard, American, 6; English, 6
Young's experiment, 207

Zero, absolute, 94, 104, 105

QC Bennett, Clarence E.
23   (Clarence Edwin),
B438 1902-
1970
      Physics without
        mathematics

© THE BAKER & TAYLOR CO.